797,885 Books
are available to read at

www.ForgottenBooks.com

Forgotten Books' App
Available for mobile, tablet & eReader

ISBN 978-1-333-71737-7
PIBN 10538725

This book is a reproduction of an important historical work. Forgotten Books uses state-of-the-art technology to digitally reconstruct the work, preserving the original format whilst repairing imperfections present in the aged copy. In rare cases, an imperfection in the original, such as a blemish or missing page, may be replicated in our edition. We do, however, repair the vast majority of imperfections successfully; any imperfections that remain are intentionally left to preserve the state of such historical works.

Forgotten Books is a registered trademark of FB &c Ltd.
Copyright © 2015 FB &c Ltd.
FB &c Ltd, Dalton House, 60 Windsor Avenue, London, SW19 2RR.
Company number 08720141. Registered in England and Wales.

For support please visit www.forgottenbooks.com

1 MONTH OF FREE READING

at

www.ForgottenBooks.com

By purchasing this book you are eligible for one month membership to ForgottenBooks.com, giving you unlimited access to our entire collection of over 700,000 titles via our web site and mobile apps.

To claim your free month visit:

www.forgottenbooks.com/free538725

* Offer is valid for 45 days from date of purchase. Terms and conditions apply.

English
Français
Deutsche
Italiano
Español
Português

www.forgottenbooks.com

Mythology Photography **Fiction**
Fishing Christianity **Art** Cooking
Essays Buddhism Freemasonry
Medicine **Biology** Music **Ancient Egypt** Evolution Carpentry Physics
Dance Geology **Mathematics** Fitness
Shakespeare **Folklore** Yoga Marketing
Confidence Immortality Biographies
Poetry **Psychology** Witchcraft
Electronics Chemistry History **Law**
Accounting **Philosophy** Anthropology
Alchemy Drama Quantum Mechanics
Atheism Sexual Health **Ancient History**
Entrepreneurship Languages Sport
Paleontology Needlework Islam
Metaphysics Investment Archaeology
Parenting Statistics Criminology
Motivational

Elementary Turning

FOR USE IN

Manual Training Classes

BY

FRANK HENRY SELDEN

FULLY ILLUSTRATED

Chicago — New York
RAND, McNALLY & CO., PUBLISHERS

Copyright, 1907, by Rand, McNally & Co.

PREFACE

THE series of exercises given in this text is the result of the author's experience in teaching turning. Each model has been developed for the purpose of teaching a correct use of the tools, so that pupils can do excellent work without the long drill to acquire skill or the necessity of scraping where cutting tools should be used. If turning lathes are to be used in the school, they should be used properly. It is the hope of the author that this manual will aid such instructors as are trying to teach a rational method of turning.

A careful examination of the text by one who understands this line of work will reveal the fact that the elementary principles are covered very completely, and yet there is not in the regular set a single exercise which may be dispensed with, without a real loss to the average pupil.

The numerous illustrations are not only to make clear a way in which to do the work, but to furnish such a variety of similar views that the pupil will be certain to draw comparisons and to form an individual method of work.

Although a proper study of this book will result in a marked degree of proficiency in turning, yet the greater benefit will be the training which comes from the constant and careful attention required to do this work. The aim is not technique, but power — mental growth.

Several of the models were suggested by those used in other schools. The general arrangement and method of treatment are entirely original with the author.

But few woods are mentioned in the text. In fancy turning a variety of woods should be used, if they can be obtained. The instructor should see that each pupil acquires some knowledge of both local and foreign woods.

CONTENTS

		PAGE
Preface	- - - - - - - -	5

PART I

Introduction	- - - - - - - -	11
Equipment	- - - - - - - -	14
Regulations	- - - - - - - -	15
Lesson I.	Placing Work in the Lathe	19
Lesson II.	Cylinder - - - - -	25
Lesson III.	Stepped Cylinder - - - -	32
Lesson IV.	Left-hand Semi-bead - - -	35
Lesson V.	Right-hand Semi-bead -	37
Lesson VI.	Half-inch Left-hand Semi-bead	41
Lesson VII.	Half-inch Right-hand Semi-bead	44
Lesson VIII.	One-inch Bead - - - -	45
Lesson IX.	Half-inch Bead - - - -	47
Lesson X.	Three-eighths-inch Bead	49
Lesson XI.	One-inch Cove - - - -	50
Lesson XII.	Three-fourths-inch Cove	55
Lesson XIII.	Half-inch Cove - - - -	57
Lesson XIV.	Three-eighths-inch Cove	58
Lesson XV.	One-inch Bead and Cove	58
Lesson XVI.	Half-inch Bead and Cove	61
Lesson XVII.	Spindle with Cones - - -	62
Lesson XVIII.	Sandpapering - - - - -	65
Lesson XIX.	Shellacing - - - - -	67
Lesson XX.	Beaded Spindle - - - -	69
Lesson XXI.	Polishing - - - - -	71
Lesson XXII.	Square-end Spindle - - -	75
Lesson XXIII.	Curved Spindle - - - -	81
Lesson XXIV.	Tapered Spindle - - - -	82
Lesson XXV.	Porch Spindle - - - -	84

8 ELEMENTARY TURNING

		PAGE
Lesson XXVI.	Plain Box	86
Lesson XXVII.	Box with Knob	92
Lesson XXVIII.	Plain Goblet	93
Lesson XXIX.	Goblet with Rings	97
Lesson XXX.	Rosette	101

PART II

SUPPLEMENTARY EXERCISES

Introduction		105
No. I.	Tool Handle	105
No. II.	Gavel	109
No. III.	Gavel Patterns	112
No. IV.	Carpenter's Mallet	113
No. V.	Carver's Mallet	115
No. VI.	Molder's Rammer	116
No. VII.	Darning Ball and Darning Hemisphere	117
No. VIII.	Glove Mender	118
No. IX.	Plain Ring	119
No. X.	Napkin Ring, First Method	121
No. XI.	Napkin Ring, Second Method	124
No. XII.	Vise Handle	125
No. XIII.	Wooden Screws	127
No. XIV.	Large Box	129
No. XV.	Box Designs	132
No. XVI.	Candlesticks	133
No. XVII.	Designs for Candlesticks	135
No. XVIII.	Hat Rests	137
No. XIX.	Combining of Woods	141
No. XX.	Designs for Goblets	143
No. XXI.	Knife and Fork Rest	144
No. XXII.	Pin Tray	145
No. XXIII.	Turned Frames	147
No. XXIV.	Chair Legs	153
No. XXV.	Chair Rungs and Spindles	156
No. XXVI.	Footstool Leg	157
No. XXVII.	Designs for Footstool Legs	158

		PAGE
No. XXVIII.	Footstool	160
No. XXIX.	Piano Stool	161
No. XXX.	Turned Stool	162
No. XXXI.	Group of Fancy Turnings	163
No. XXXII.	Turned Molding	163

PART III
TOOLS AND FITTINGS

Introduction	167
Arbors	168
Calipers	168
Chisels	170
Chucks	172
Screw Chuck, See Chucks	173
Spur Chuck, See Chucks	177
Compasses	178
Dead-center	178
Face-plate	179
Gauges	179
Gouges	181
Lathes	184
Oilstones	187
Parting Tool	187
Scraping Tools	188
Ring Tools, See Scraping Tools	189
Live-center, See Spur Center	190
Sizing Tool	190
Spur Center	190
Templet	191

ELEMENTARY TURNING

PART I

INTRODUCTION

This course in turning is intended to give elementary exercises only. Each model in Part I is given with a definite purpose and should not be omitted. There will be little need for class demonstration. Each pupil should have a book at his bench, and should take it home with him often enough to gain in advance a definite idea of each day's lesson.

Each piece should be turned with care and in the order given, and the exercise should not be repeated. No matter what the plans of the pupils may be, much time will be saved by making each of the twenty models before attempting any fancy turning. If the first eight or sixteen pieces have been made, and the work is very poor, it is better to return to the first piece and begin again. Thus continuing the study of principles, rather than acquiring skill to do the work by mere repetition.

Always keep in mind that turning cannot be done with dull tools. Do not resort to scraping the pieces where they should be turned. Do not use any sandpaper until the fifteenth exercise, and then use only *No.* ½. Although this set may appear to consist of too large a number of pieces, a proper use of them will demonstrate that they are a much shorter and quicker road to successful turning than the less numerous exercises given by others.

The methods of using tools in turning on modern lathes and with modern tools vary somewhat from the methods used when lathes were more cumbersome and tools not so easily obtained. In the school shop such tools and methods should be employed as will tend most to an active mental direction of the process, and give as little occasion as possible for the acquiring of skill.

The work must be carried on in such a manner that there is a continuous increase in power of attention and ability to do a given amount of work in a definite period of time. The pupil must learn to keep up with the lathe, and this by developing power to think more quickly and accurately, rather than by acquiring skill. If the work is planned to develop skill, the result will be injurious rather than helpful.

The material for the first exercises should be of pine, because it is easy to work, when the tools are used properly; and because any attempt to scrape the piece to shape is easily detected.

One of the first things to decide in learning to use the lathe is whether one shall turn right-handed or left-handed. Either way is easily learned, whether one is right-handed or not. But, when the decision is once made, do not change. The righthand position is probably the better for a large proportion of work, although the left-hand position seems to be easier in some of the first exercises.

As a large part of the time is necessarily taken up in learning the use of the tools, you will have but little time for fancy turning, unless you are very careful to learn the correct use of your tools. You should be especially careful in turning the first pieces, for the more nearly correct you use your tools in the beginning, the more rapid will be the progress and the better will be your work.

The first exercises are so designed that, if properly used, they will readily give a freedom and certainty which is not the result of skill, but of an exact understanding of the process. After this knowledge has been acquired, a great variety of articles may be made in a short space of time.

If you learn the correct use of each turning tool, you will be able to turn fancy articles of knotty, hard, or cross-grained wood. Such wood is often much more beautiful than that which is plain and straight-grained. This ability to use the tools will not be lost, even though you do no turning for a considerable length of time.

EQUIPMENT

PERSONAL EQUIPMENT

Each pupil must provide himself with a pocket-rule, two-foot, four-fold; a lead pencil, one combination India oilstone, one hard Arkansas oilstone slip.

The apron used in joinery may be used. A jumper should also be worn.

SCHOOL EQUIPMENT

Each drawer is provided with three skew chisels (1-inch, $\frac{1}{2}$-inch, and $\frac{1}{4}$-inch), three turning gouges ($\frac{1}{2}$-inch, $\frac{3}{8}$-inch, and $\frac{1}{4}$-inch), a $\frac{1}{2}$-inch round nosed scraping tool, a 1-inch firmer gouge for roughing and a $\frac{1}{8}$-inch parting tool. Gouges for heavy work, and special tools for rings, etc., are provided in the tool room.

STOCK

In both Part I and Part II the stock used in every case, where allowable, is of the same size. This avoids much waste time, which would occur if a variety of sizes were used. Stock 8 inches long by $1\frac{3}{4}$ inches square appears to be the best size for exercise pieces and also for small footstool legs.

Goblets, napkin rings, and similar objects may be made from the short pieces resulting from the cutting of regular stock. Chair legs, large footstool legs, candlestick stems, etc., should be selected from the better portion of the $1\frac{3}{4}$-inch stock.

REGULATIONS

In the lathe room, while the lathes are in motion, there is always a probability that work will be injured if from any cause a pupil looks up while his tools are cutting. It is, therefore, a matter of much importance that pupils should refrain from all conversation, and from moving about the room. Care should also be taken to avoid any unusual noise in turning, or in starting or stopping the lathe.

No pupil should ever borrow or lend any tool or piece of material. Every piece of material, including sandpaper, should be plainly marked with the pupil's name. When the work is completed, the name of the pupil and the date of completion should be plainly written upon it. The work should be kept in the bench drawer until completed.

The lathe should be watched, and any indication of its being out of order should at once be reported. A drop of oil should be placed on each bearing of the live spindle at the beginning of each recitation. The end of the piece bearing against the dead-center should be oiled when the piece is first placed in the lathe, and each succeeding day that the same piece is used. The dead-center should be carefully watched, and, if it becomes too warm, the tail-screw should be turned to loosen the work, or more oil be applied. In case **any** tool or bit of material has been tampered with during the absence of the pupil, it should be reported to the instructor at once.

At the close of the recitation the tools must be put in place. Tools which require grinding may be handed to the instructor. Each pupil must brush all shavings and dirt from his lathe; and when cleaning the lathe care must be taken that no dirt is thrown on adjoining lathes. The lavatories are for use, and every pupil should wash his hands and brush his clothes before going to another recitation.

At the close of the year each pupil may remove the work he has completed by paying for the materials used, except such pieces as are needed in the school for exhibition.

For each exercise a sufficient amount of material will be given each pupil. This material will be sufficient to complete the exercise properly, and only in very extreme cases shall more material be given. The first piece given must be finished as well as possible, even though very incorrect or under size. Sandpaper should not be used on any exercise until that exercise has been passed upon by the instructor.

THE ILLUSTRATIONS

The illustrations for this book represent the actual conditions and work of a school room where pupils succeed in learning to use their turning tools, as they are used by good workmen in practical

ELEMENTARY TURNING 17

turning. A large number of pupils were asked to pose for the views, in order to eliminate as much as possible the peculiarities of any one pupil, and illustrate general principles, applicable to all.

In studying the illustrations, do not attempt to imitate them, but rather follow the principles given, adapting them to your own strength and temperament. All important positions are shown from different sides and by different pupils. Examine all illustrations relating to the exercise before attempting to do any turning. This will lessen any liability to misunderstand the illustrations.

If you have already formed habits in the use of turning tools, do not continue them if there is a better method. Often there are several ways which are correct, but this does not imply that any method will answer. Learn the best way, as it will save you much time and trouble.

In many of the illustrations a part of the lathe centers are shown. This is to indicate the position of the piece in the lathe. It is a matter of considerable importance which end of a piece is on the live-center, and whether there is a stub at either end to be cut off after the work has been removed from the lathe.

LESSON I

PLACING WORK IN THE LATHE

There are several good methods of centering pieces which are to be placed between the lathe centers. A method seldom used is to draw diagonals across the end, as shown in Fig. 2, and also in Figs. 230 and 259.

If a common marking gauge is at hand, it may be set for a space a little less than half the width or thickness of the piece, and four lines drawn, making a small rectangle or square at the center.

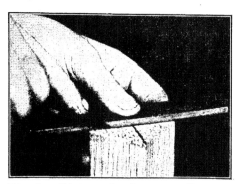

Fig. 2. Finding the Center by Drawing Diagonals.

Another method is to draw four lines on the end with the compasses, as shown in Fig. 3. This is a very good way to do, and it is used by many turners. Care must be taken to have the end of one leg of the compasses against the bench, as the other leg draws the line. Hold the compasses so that one leg will be exactly above the other leg.

Perhaps the best way in which to find the center is to lay a piece on the bench, and draw lines by moving a pencil along the top edge of the strip and against the piece to be centered, as shown in Fig. 4. The rule may be used for this purpose. It sometimes happens that the cleat on the bench-hook is just the correct thickness for use in centering.

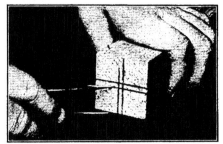

Fig. 3. *Using the Compasses to Find the Center.*

After the piece has been properly centered, place one end against the live-center, the left hand holding the end nearest to the dead-center (Fig. 5). Then grasp the hand wheel, and turn the screw in the tailstock until both centers have been forced into the piece sufficiently to hold it securely, while it revolves

Fig. 4. *Finding the Center with Pencil and Strip.*

ELEMENTARY TURNING

against the tools (Fig. 6). Next loosen the tail-screw so that you can put a little oil into the depression made by the dead-center (Fig. 7). The oiling must not be done while the lathe is in motion.

Fig. 5. *Placing a Piece Between the Lathe Centers.*

Fig. 6. *Tightening the Tail-screw.*

Retighten the tail-screw, making it as tight as it can be, and allow the spindle to revolve freely. By placing the hand on the cone pulley, as in Fig. 10, and revolving the live-center while adjusting the tail-screw, the proper tension can be determined. After the dead-center has been adjusted, turn the clamp screw handle,

Fig. 7. *Oiling the Dead-center.*

H, Fig. 8, until it is tight. This is to hinder the dead-center from moving away from the work.

Fig. 8. Tightening the Clamp Screw.

Do not use a mallet to drive the piece on to the live-center, for it is quite as essential that the dead-center form a good bearing, as that the live-center be forced into the piece

Revolve the piece until a line through the two opposite corners will be horizontal. Loosen the set screw, A, Fig. 9, and adjust the tee rest until the

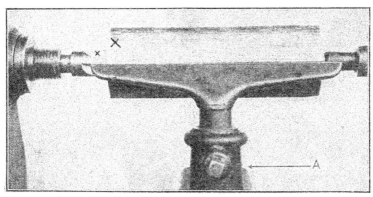

Fig. 9. Setting the Tee Rest.

top is on a level with this line. Use the rest at the same height for all turning similar to the twenty exercises. Very tall pupils may use the rest a little above the center, and very short pupils may use the rest a little below the center. Each one should determine at the beginning the proper height, and not change it for any of the twenty exercises. It should remain at the same height for both skew chisel and gouge turning.

The rest should be as close to the wood as will allow the piece to revolve. After you have become familiar with the use of the tools, the rest need not be moved up to the piece after it has been turned to a cylindrical form; but in turning the first exercises, it will probably be better to move the rest close up to the piece, as soon as it has been made cylindrical. Sometimes the ends will remain square, and you will be obliged to turn while the rest is at some distance from the part you are cutting.

In advanced turning the rest will need to be adjusted to a variety of positions; and in some cases the height will have to be changed, but for all ordinary turning the rest should be kept at the same height.

Before placing any piece of wood in the lathe it should be carefully examined. Small defects, such as worm holes and sap, need not be considered in the first exercises. Small knots are usually not difficult to work in the lathe, and pieces containing them should not be discarded. Wain at corners does no harm, if it is not so large that the blank will not form a cylinder the full size of the piece.

The one defect that must be carefully watched is shake. Sometimes pieces, which at a glance appear all right, on close inspection will be found to contain latent checks which render them unfit for turning. These seams or shakes sometimes allow the piece to separate as it revolves, spoiling the exercise, and wasting time. Usually, the best end of the blank should be placed on the live-center, as this center tends to split the wood.

Before removing the piece from the lathe, make a pencil mark, as shown in Figs. 9 and 10, on the end to correspond with the mark on the live-center, so that after the piece has been removed from the lathe, it can be replaced exactly in the same position.

Before starting the lathe, examine the belt to see on which step of the cone pulley it is running. For these exercises it should be on a step that will give about 3,500 revolutions per minute. Larger work should not be revolved so rapidly. If the piece is of an irregular shape, it should not be revolved at so great a speed until it has been turned down to a cylindrical form.

CAUTION

If you wish to stop the lathe, do not do so by grasping the work, but place your hand on the cone pulley, after the belt has been shifted to the loose pulley. If, for any reason, you wish to touch the work while it is revolving, bend your finger, as is

ELEMENTARY TURNING 25

shown in Fig. 10, and allow only the end of your finger to touch the piece. To grasp the piece, even though it may be quite smooth, is not the proper thing to do. Do not attempt to touch it on either the upper or the lower side, but always use the end of one finger against the back side.

Fig. 10. *Stopping the Lathe and Testing the Surface.*

LESSON II

CYLINDER

The stock for this and the nineteen following exercises should be 8 inches long by 1¾ inches square. This exercise is to teach the use of the roughing

Fig. 11. *The Cylinder.*

gouge, and some of the uses of the skew chisel. Be sure to have in mind what is said in Lesson 1 about putting the work in the lathe and adjusting the rest, etc. After you are certain that everything is all right, slowly shift the belt so that the piece will revolve. Fig. 184 shows the left hand grasping the belt shifter.

Take the position shown in Fig. 12, with the hands and roughing gouge held as in Fig. 13 or 14. Whether the position taken is similar to that in Fig. 13 or 14 is not important, In the latter, the hand is turned to hinder the shavings from striking the face. By comparing Figs. 20, 41, 44, 47, 53, etc., it will be observed that the fingers of the left hand are used in a variety of positions. This is because

Fig. 12. Position While Using a Roughing Gouge.

ELEMENTARY TURNING 27

the jar of the lathe tends to numb and tire them if they are used long in exactly the same position.

The points to be kept in mind are: First, the tools must be firmly held; second, some part of the hand or fingers should come in contact with the rest; third, the angle should be such that the tools will cut rather than scrape; fourth, the tool should be firmly held upon the tee rest, and also upon the piece which is being turned; fifth, the angle which the tool makes with the line of the centers is very important, and must be carefully determined for each tool and each piece of work.

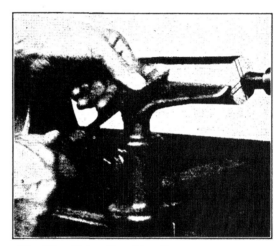
Fig. 13. Using Roughing Gouge.

By comparing Figs. 13 and 14 you will see that the roughing gouge is held at right angles to the centers, and at as oblique an angle vertically as will allow the cutting edge to enter the wood. Be sure that your gouge is sharp. Read carefully what is said in Part III in regard to sharpening gouges.

Do not attempt to cut the piece rapidly, but rather see how fine and how even you can cut the shavings. Move the gouge the entire length of the piece. If the piece were longer, you would turn

Fig. 14. Hand Shielding Shavings from the Face.

down a place at one end, and then little by little work towards the other end, finishing a small part of the surface each time you move the gouge from left to right, as in turning the table leg (Fig. 237).

ELEMENTARY TURNING

After you have cut off a little of the piece, stop the lathe, as shown in Fig. 10, and examine the work to see how much has been cut away, and whether the gouge is cutting smoothly or tearing the surface. Study the positions of the tools in Figs. 12, 13, 14, 100, and 113.

If the roughing gouge is held properly, it will cut quite smoothly, as the shape of the end of the gouge is such that a shaving is cut. Should you use a turning gouge for roughing, you would discover that it does not cut as freely nor as rapidly, and, hence, the common firmer gouge is used in turning as a roughing gouge; or else a turning gouge is ground like a firmer gouge.

In days gone by, when tools were more expensive and labor cheaper, the turner used as few tools as possible, and therefore used his large turning gouge for roughing. At present such a use of the turning gouge must be considered very much out of place. Should the wood to be turned be so rough or knotty that the light gouge might be broken, it would be proper to use the heavy turning gouge.

In roughing the edges of pieces on the faceplate, as shown in Figs. 124 and 125, the turning gouge is always used. Continue using the gouge until the piece is cylindrical the entire length. The gouge will not produce a straight finished surface no matter how carefully used. To give the work the even, glossy surface, a turner's skew chisel is required.

For smoothing work of this size, a 1-inch skew chisel may be used. It should be held as shown in Fig. 15, 16, or 17. Each of these views shows the skew chisel held at a slightly different angle. Also see Figs. 27 and 29.

When the skew chisel is held as shown in Fig. 16 or 17, the point is not as liable to catch and injure the work, but it will not cut so smoothly, and will dull much more rapidly, especially if the piece is cross-grained. Begin by holding it so that the cutting edge is

Fig. 15. Smoothing a Cylinder (See Figs 16 and 17).

Fig. 16. Using a Skew Chisel.

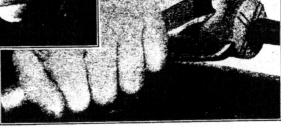

Fig. 17. Using a Skew Chisel (See Figs. 15 and 16).

ELEMENTARY TURNING 31

at quite an angle to the center line of the piece (Fig. 16 or 17), and gradually change the position at which you hold it, until the cutting edge is nearly parallel to the center line (Fig. 15).

Observe carefully that the skew chisel is held as shown in Fig. 16 while cutting toward the left, and as shown in Fig. 15 or 17 while cutting toward the right. It is very important that you change the position of the chisel in this manner, for it not only rests upon the tee rest, but also upon the piece being turned; and if you attempt to cut at the end of the piece with the chisel, unsupported by the wood, it will be quite sure to go deeper than you wish, and may spoil the piece.

Try to make the surface of the cylinder smooth, practicing near the right-hand end. Then smooth a space about $1\frac{1}{2}$ inches long at the left-hand end. As this is your first piece, you will probably not be able to make the piece smooth and straight its entire length, but you should make it quite smooth near the left-hand end. Do not attempt to smooth the ends of this, or any other piece used for the first twenty exercises. On pieces of this character the ends are not usually smoothed; and if they were to be smoothed, the operation would be found to be quite difficult. Remember that none of the first sixteen exercise pieces are to be sandpapered.

ELEMENTARY TURNING

LESSON III

STEPPED CYLINDER

Use the piece worked to a cylinder in Lesson 2. Set the rest close to the piece, and with the rule and

Fig. 18. *Stepped Cylinder*.

pencil, as shown in Fig. 19, make a mark, while the lathe is in motion, one inch from the left-hand end. With the acute point of the skew chisel cut a small groove at the place marked by the pencil. Hold the skew chisel as shown in Fig. 39.

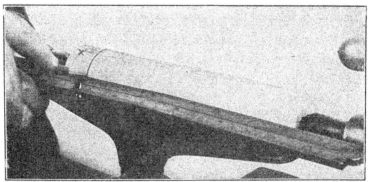

Fig. 19. *Marking Spaces with a Pencil*.

These inch spaces may be marked by using the chisel, as shown in Fig. 31, instead of the pencil.

ELEMENTARY TURNING 33

The pencil is better and easier at first. After you have become more familiar with the lathe and tools, you can use other methods for marking spaces.

After you have made a mark and cut a groove with the skew chisel, remove about $\frac{1}{16}$ inch in thickness of material from the right-hand end of the piece up to the mark. Do not attempt to turn off this waste with the skew chisel, but use the roughing gouge, holding it as shown in Fig. 13, until it is near the groove; then roll the gouge so that the corner will cut close to the shoulder, as shown in Fig. 20.

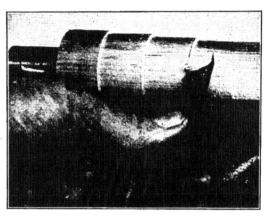

Fig. 20. Rolling the Roughing Gouge.

After the roughing gouge has been used, hold the skew chisel as shown in Fig. 15, 16, or 17, and smooth the smaller part of the cylinder in the same manner as described in Lesson 2 for making a cylinder.

In order to true the surface close up to the shoulder, the handle of the skew chisel should be lowered until the obtuse corner of the cutting edge completes the cut (Fig. 21). The tendency is to roll the handle instead of lowering it. This should not be done for it is quite certain to cause the corner to cut too deeply. Should you wish to smooth the surface to the right hand, do not forget to change the skew chisel to the position shown in Fig. 15. While smoothing towards the left hand of each part, hold the chisel as shown in Fig. 16. Continue marking off inch spaces and cutting steps, until the piece has the shape shown in Fig. 18

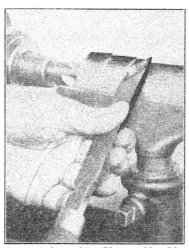

Fig. 21. Smoothing Up to a Shoulder.

It is not so essential that each step be exactly $\frac{1}{16}$ inch, as that the surface between the steps be exactly straight and smooth. Should you be obliged to make the large ends less than $1\frac{3}{4}$ inches in diameter, the steps may be only $\frac{1}{32}$ inch. Be careful to cut no deeper with the point of the skew chisel than the amount of the step, for any mark at this place left in the finished piece shows badly.

Before presenting the piece for inspection, write your name and the date on the surface, near the large end. Do not forget to mark the end so that it can be replaced in the same position on the live-center.

LESSON IV

LEFT-HAND SEMI-BEAD

This exercise is turned in the same way as the stepped cylinder, and then the curves are cut to the shape shown in Fig. 22. To work these curves, the skew chisel is laid on the piece the same as in smoothing a cylinder (Fig. 16). The right hand is then raised in a curve so that the chisel cuts a little nearer the obtuse angle as it approaches the inner end of

Fig. 22. Left-hand Semi-bead.

the curve (Fig. 23). As the movement is finished, the chisel cuts at the extreme obtuse corner, and instead of a shaving being cut, a small ring is formed, which breaks in two as it is crowded against the square end of the adjoining semi-bead.

Do not attempt to cut thick shavings, but proportion the material so that each shaving will be of sufficient size to cut easily. Try to take the last shaving from the whole surface being shaped.

There is always danger of resting the skew chisel on the work so heavily that it will follow the grain of the wood, and the piece be turned out of round.

36   ELEMENTARY TURNING

In case the skew chisel is jarred by the revolving of the piece, bear more firmly upon the rest. Sometimes this jarring is stopped by holding the chisel more nearly straight with the work. Compare Fig. 15 with Figs. 16 and 17.

When the piece is cross-grained, the skew chisel must be held nearly straight, as shown in Fig. 15.

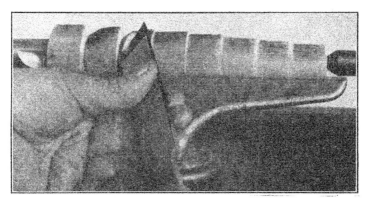

Fig. 23.   Turning a Curve.

If your tools are properly sharpened and correctly used, the work will be quite smooth, even though the piece is cross-grained.

It is not so much practice to gain skill, as it is a careful study to gain a correct knowledge of the proper methods of using the tools that will give success in this work.

If the chisel catches, do not think that it is because of the grain of the wood or because the chisel is not held with sufficient force, for it is quite

ELEMENTARY TURNING 37

probable that the cause of the trouble is the angle at which the chisel has been held, or you have been cutting too near the point.

It is not a difficult matter to turn these curves by holding the chisel with only the right hand, as shown in Fig. 24; therefore, study and use the correct movements. This figure shows the chisel in the extreme position.

Fig. 24. Turning with One Hand.

Usually the handle would not be raised so high, or the hand moved so far to the right.

Notice that the top of the rest remains on a level with the lathe centers.

LESSON V

RIGHT-HAND SEMI-BEAD

This exercise is the same as Lesson 4, except that the curves are in the opposite direction, and the last division is omitted to avoid hitting the live-center.

If in turning the second exercise the skew chisel was held too close to the body, you will now have trouble with the chisel catching, because you will raise it directly up, or simply revolve it, instead of moving the handle in a curve toward the left. In either case the skew chisel will frequently catch.

By comparing Fig. 24 with Fig. 26 you will see how the positions differ in working the two curves. These pictures, although showing the angle at which the chisel should be held, if but one hand were used, show the real principle of changing the angle of the skew chisel in working the two sides of a bead. Fig.

Fig. 25. Right-hand Semi-bead.

27 is the same as Fig. 26, excepting that both hands are used, and the chisel has not passed to the extreme position.

Do not simply change your position so that you will be able to turn this exercise, and then take a different position for turning curves in the opposite direction, but learn to reach out far enough to turn curves either way without changing the position of your feet on the floor. If this is not done, there will be much trouble when an attempt is made to turn complete beads.

In all these exercises and all similar light work, the arms should be free, never resting against the hip or side.

ELEMENTARY TURNING 39

In turning very heavy work, sometimes the arm must be held against the side in order to hold the tool steady. Such work, however, does not require so frequent a change of position, and the

Fig. 26. Turning with One Hand.

workman can step about the lathe as often as required to bring his side in line with the handle of the tool.

If the tools catch and the piece is spoiled, the exercise should not be repeated. You should pass on to the next exercise. By attempting the next problem you will have an opportunity to see the

Fig. 27. Using Skew Chisel.

ELEMENTARY TURNING

same difficulty from a different view point, and it may enable you to overcome it. Only by a study of the methods, instead of blindly practicing for skill, will you become able to do good and rapid work. To repeat an exercise simply to gain skill cannot result in a knowledge of turning, and is certain to injure the mind, although by such repetition you may be able to do some very good work.

LESSON VI

HALF-INCH LEFT-HAND SEMI-BEAD

This piece should be of the same diameter at both ends. It is not necessary to caliper it, but judge its size carefully by examining it with your

Fig. 28. Half-inch Left-hand Semi-bead.

eye only. A more accurate judgment may be made if the piece is removed from the lathe, and held up to the light. Of course, you should try to judge correctly while the piece is in the lathe, and with patient effort you will be able to do so well that you will seldom have to remove a piece from the lathe for examination.

42   ELEMENTARY TURNING

Before attempting to mark the spaces for the curves, smooth the entire surface with the skew

Fig. 29. *Using the Skew Chisel.*

ELEMENTARY TURNING 43

chisel. As the piece in this exercise is the same size throughout, and the spaces are but one-half inch, the curves should all be alike, *i. e.*, they should each be the shape of a quarter circle.

Now that you have turned both right-hand and left-hand curves, you should be able to turn these without holding the chisel close to your body. Stand in such a position as will allow of turning either right-hand or left-hand curves without changing the position of your body. Fig. 29 shows very clearly how such a position is taken.

It frequently happens in turning the curves that the square end of the adjoining bead is roughened. This end should be cut smooth before the piece is considered finished. To do this, hold the skew chisel as in cutting at the sides of the coves (Fig. 39), turning the handle to the right or to the left as required to give the proper angle to the end, but do not tip or roll the tool out of the vertical position. If the skew chisel, when used in such a place, is revolved so that it cuts at a place on the edge above the point, it is almost certain to catch.

After you have done your best to work each curve properly and they are not satisfactory, use the skew chisel, as shown in Fig. 16 or 17, and flatten each curve a little, so that you will have a flat space on which to lay the end of the chisel in re-cutting the curves. Be very careful to keep each part of correct size. Although size is not the most impor-

44 ELEMENTARY TURNING

tant part of the exercise, yet you should begin at once to work as nearly to size as you can.

LESSON VII

HALF-INCH RIGHT-HAND SEMI-BEAD

This is the same as Lesson 6, except that the curve is in the opposite direction. The same diffi-

Fig. 30   Half-inch Right-hand Semi-bead.

culties are met as in Lesson 5. Great care should be taken in this exercise to make the curve a correct quarter circle.

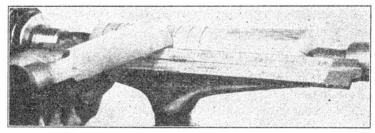

Fig. 31.   Marking Spaces with Skew Chisel.

You should now be able to use the skew chisel with sufficient accuracy to mark the spaces, as shown in Fig. 31.

ELEMENTARY TURNING 45

In using the point of the skew chisel in this manner, make but a very light mark. If you wish the point to cut deeper, go over the lines again after the rule has been removed.

A very deep cut cannot be made except by cutting out a bit of the material, for if the point of the chisel is held long in one place or pressed hard into the wood, the friction caused will heat the point and color it and may injure it very much. The tools should not be held so hard or so long against the wood as to color them even at the extreme thin edges or points.

Try to make the curves so even that there will be no mark showing where the skew chisel began to cut. Also be careful not to rub the skew chisel on the piece so hard that the grain of the wood will be bruised or crushed. Keep in mind that in proper turning the tools must cut evenly and smoothly, and that the surface must be glossy. When you have acquired the correct way of handling your tools, you will soon be able to work with considerable speed.

In case the curves are not of correct form, cut them down a little with the roughing gouge. Smooth these places with the skew chisel, then try again to work the curves.

Be careful to keep the spaces equal. After cutting the curves part way down, test the spaces with the rule and pencil, as shown in Fig. 19.

LESSON VIII

ONE-INCH BEAD

Be careful to work this piece to correct size. If a lead pencil line is made at the center of each bead

Fig. 32. One-inch Bead.

(Fig. 32), the turning of the bead is apt to be more satisfactory.

Work carefully, leaving the pencil marks to be seen when the work is finished. Avoid cutting too deep between the beads as you mark the spaces with the point of the chisel, and also as you turn the beads.

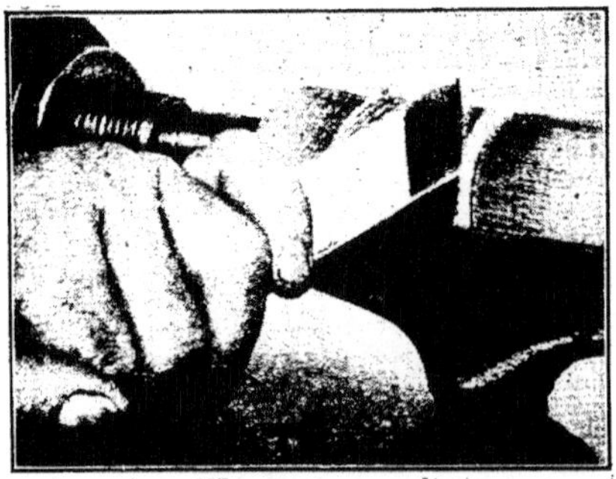

Turn from two adjoining beads down to the central space, taking a shaving first from one bead and then from the other (Fig. 33). This is much better than turning both sides of a bead at once, as this

Fig. 33. Turning a Bead.

ELEMENTARY TURNING

requires the making of a square shoulder at the adjoining bead.

When the turning is finished, there should be no unevenness between the beads. The curves of the two adjoining beads should exactly meet. Any roughness at this point hinders the giving of a proper finish to the piece. Even though the piece is not to be sandpapered or shellaced, the work should be done as if it were to be finished in this manner.

LESSON IX

HALF-INCH BEAD

This exercise is based on the same principle as is given for the one-inch bead. The curves are

Fig. 34. Half-inch Bead.

somewhat steeper, and should be turned with greater care. These beads should be turned evenly, and without any roughness or mark between them.

When the exercise is finished, the beads should be of the same size. If some are of greater diameter than others, cut them down, but do not cut them entirely off. Then try again.

There is quite a tendency in making these beads, to make some much wider than others. Test them often with the rule. If you are careful to cut a shaving of equal thickness from each side of the line, the beads will be of the same size.

As the beads decrease in size, the swinging of the chisel by the right hand is less, yet you must not hold it in the same line and simply turn it. The large curves are given to teach this swinging motion of the right hand, and if you are to become able to do good and rapid turning, you must continue to use this motion. It is the same in kind, though differing in degree, no matter what size of curve you are turning. Without this swinging motion or changing of the angle which the tool makes to the line of the lathe centers, the end of the chisel is not properly balanced between the work and the shaving, and must catch and injure the piece, unless the chisel is held in position by main force. To apply so much strength is often quite difficult or impossible. Even if you are able to do so on these practice pieces, you will find great difficulty in applying so much strength in turning things for use.

Do not make the mistake of attempting to do the turning by taking so very fine a shaving that the chisel will not catch, although it is not held at the proper angle. Such work is only a kind of scraping and can never result in doing good work or in learning to turn.

ELEMENTARY TURNING 49

The peculiar conditions of grain, etc., are certain to make trouble for you, unless you learn the correct method of using the chisel. When the chisel is used correctly, you will find it quite an easy matter to turn cross-grain, knots, and almost any sort of a piece.

LESSON X

THREE-EIGHTHS-INCH BEAD

Fig. 35. *Three-eighths-inch Bead.*

This exercise is worked the same as the 1-inch beads and the ½-inch beads. The ⅜-inch beads may

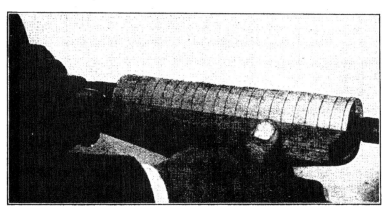

Fig. 36. *Using a Gauge Stick.*

be cut with the ½-inch skew chisel. This size bead is probably used more than any other in the regular turnings for stair and porch work.

To make all the divisions exactly ⅜ inch is quite difficult. To assist in making these divisions, you may use a gauge stick, as shown in Fig. 36. Hold the stick firmly, and unless it fits the piece exactly, press it against one end, and gradually change the pressure until it marks throughout the length of the piece. Hold the stick so that the spurs point towards the axis line of the cylinder. The making of a gauge stick is described in Part 3.

LESSON XI

ONE-INCH COVE

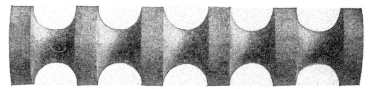

Fig. 37. One-inch Cove.

This exercise introduces the turning gouge. For directions for grinding and whetting gouges see Part 3. Be sure that the gouge is in proper condition before attempting to use it.

Turn the piece to a smooth, straight cylinder. Lay off the spaces as indicated by the drawing

ELEMENTARY TURNING 51

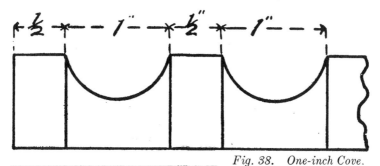

Fig. 38. One-inch Cove.

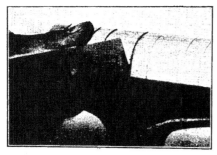

Fig. 39. Cutting with Point of Skew Chisel.

(Fig. 38). Start the coves by making cuts with the point of the skew chisel, as shown in Fig. 39.

For turning coves of this size and smaller sizes, including ½-inch, the ½-inch turning gouge should be used.

In first attempting to use the gouge for working coves, it is better to start the opening

Fig. 40. Starting Coves with Gouge.

by cutting out a small amount of waste material, as shown in Fig. 40.

The gouge is here held in the opposite position to that in which it is used while cutting the cove. This

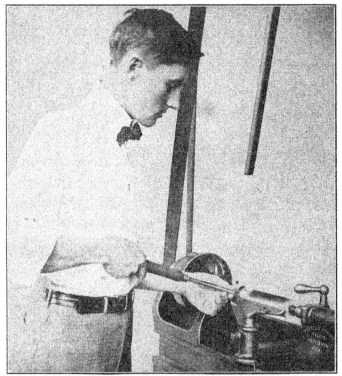

Fig. 41. *Turning a Cove.*

is to make an opening in the surface so that it will not be so difficult to keep the gouge from cutting back of the line and spoiling the work. After the space has been started in this manner, it will appear as in Fig. 40 or 42. To finish the cove, hold the gouge as shown in Figs. 41 and 42, and gradually move it forward and upward, until it has the position shown in Figs. 43 and 44. Then place the gouge at the other side of the cove, and move it in the same manner.

Continue to repeat these movements, cutting a shaving alternately from each side, and each time making the cove a little deeper, until it is of the proper depth. Always strive to have the shavings from each side meet at the center, so that there will be no unevenness where they come together. This is really the most difficult part of the work in turning coves.

Fig. 42. Turning a Cove.

If the gouge is used properly, the finished piece will have the appearance shown in Fig. 37.

If the gouge passes beyond the center it will scrape the wood instead of cut it, and will be soon dulled.

If the gouge catches in starting, it is probably caused by not holding it so that it cuts at the extreme point, as indicated by the sketch Fig. 45, and illustrated in Fig. 47.

54 ELEMENTARY TURNING

The common mistake in the use of the gouge is the failure to lower the handle so that it will cut near the top of the piece as it reaches the center of

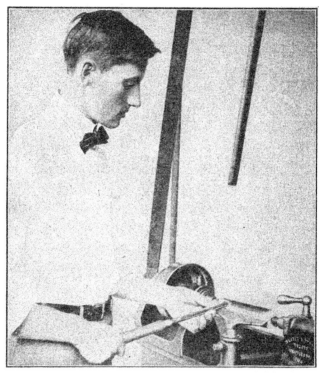

Fig. 43. Turning a Cove.

the cove. If the handle is not lowered but rolled, the gouge will scrape instead of cut. It will be quickly dulled if it scrapes, and the cove will be rough instead of smooth and bright.

ELEMENTARY TURNING 55

The correct shape of the cove may be secured by this scraping motion, but it will not be good work, and it will be easily distinguished from work which has been properly done. There is no reason why this work should be done by scraping,

Fig. 44. Turning a Cove.

as it is not a difficult matter to learn how to do it properly; and when the proper way is once learned the work can be done much faster in the right way than it can be done in the wrong way.

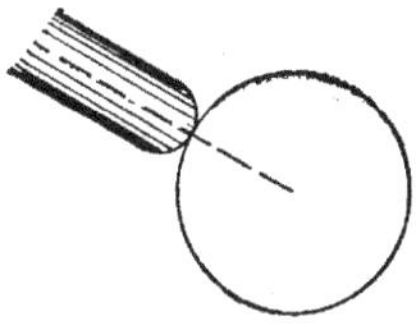

Fig. 45. Starting Gouge.

LESSON XII

THREE-FOURTHS-INCH COVE

This exercise is similar to that given in Lesson 11. The difference being that the coves are but ¾ inch. The spaces between the coves are each ½ inch. The straight parts between the coves are not worked after the piece is spaced for cutting the coves. Therefore, the cylinder should be very carefully smoothed before marking the spaces. These coves

56 ELEMENTARY TURNING

should be exactly $\frac{3}{8}$ inch deep, and each should be an exact semi-circle. It is as great a mistake to make the coves too deep as it is to leave them too

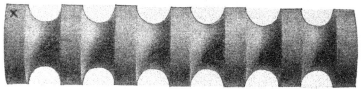

Fig. 46. *Three-fourths-inch Cove.*

shallow. One who has made the 1-inch cove correctly, as given in the previous lesson, should now be able to make these coves of correct shape and size.

You can try using the gouge without first cutting back from the line, as in Fig. 40, if you wish. The

Fig. 47. *Starting the Cove.*

proper position for starting the cut in this manner is shown in Fig. 47. The handle is moved from the upper to the lower position, the same as shown in Figs. 41 and 43. Care must be taken to keep the

ELEMENTARY TURNING

gouge from bruising the corner and making the cove too wide.

In order to avoid bruising the edges, a light cut should be taken at first.

LESSON XIII

HALF-INCH COVE

In this piece the coves and the spaces between them are each ½ inch.

The work required in making ½-inch coves is the same as given in Lessons 11 and 12 for making the

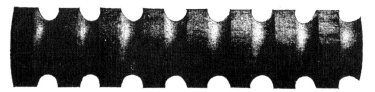

Fig. 48. Half-inch Cove.

1-inch and the ¾-inch coves. Because these curves are smaller, you will need to be more careful to make them of correct size and shape.

You should now be able to cut the curves so nicely that there will be no marks left from the point of the skew chisel at the edges of the flat parts. The bottom of each cove should be so smooth that you cannot see the point at which the shavings from each side meet.

LESSON XIV

THREE-EIGHTHS-INCH COVE

The piece for the ⅜-inch coves should be turned with the greatest care, as it is the last of the cove exercises. Be sure that the cylinder is of full size, and that it is also very smooth and straight before

Fig. 49. Three-eighths-inch Cove.

cutting the coves. The spaces between the coves are each ⅜ inch.

Some may prefer the ¼-inch gouge for this size of cove, yet a ⅜-inch gouge will do the work better if it is correctly used. This piece may be spaced with the gauge stick used for the ⅜-inch beads, as shown in Fig. 36.

LESSON XV

ONE-INCH BEAD AND COVE

Turn first the coves as indicated by the dotted lines of the drawing (Fig. 51), cutting straight down to the point where the curves will be joined when the bead is turned. Be careful to finish the coves, as shown in Fig. 52, before attempting to turn the beads.

ELEMENTARY TURNING 59

The beads may be worked either with the gouge, as shown in Fig. 53, or the chisel may be used, as in

Fig. 50. One-inch Bead and Cove.

Figs. 23, 33, and 54. Whether the skew chisel or the gouge would be used by the practical turner would be determined by circumstances. For ordinary

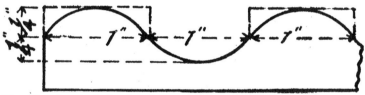

Fig. 51. One-inch Bead and Cove.

cheap work the gouge would probably be better, as it would avoid a change of tools, and the gouge can be used more rapidly. For very fine work the skew chisel must be used.

In making this piece two of the beads may be turned with the gouge, and two of them with the skew chisel.

Fig. 52. Bead and Cove, Coves Completed.

After the tools are understood, there will be many opportunities for the pupil to determine which tool to use for a certain piece of work.

If the piece is to be highly finished, use the tool that will do the smoothest work; if the finish on the

Fig. 53. *Turning Bead with Gouge.*

piece is not important, use the tool that will do the work in the least time.

Do not forget that it is a great waste of time to use a tool in such a manner as will dull it rapidly,

Fig. 54. *Turning Bead with Skew Chisel.*

even if by such a use a part of the work may be done more quickly.

One of the chief errors of this nature is the use of chisels and gouges for scraping instead of holding

ELEMENTARY TURNING

them so that they will cut shavings. Such a use of these tools dulls them very rapidly, because the wood revolves across the cutting edge, and is torn off instead of being cut.

LESSON XVI

HALF-INCH BEAD AND COVE

This exercise is similar to the previous one. The coves and beads are each $\frac{1}{2}$ inch. If the coves are

Fig. 55. Half-inch Bead and Cove.

made $\frac{1}{2}$ inch deep, the curves will all be half circles. Be careful to cut the coves to exactly the correct depth.

Fig. 56 shows the piece with the coves finished. Notice that they are straight down at each side for $\frac{1}{4}$ inch, and that the bottom is an exact $\frac{1}{2}$-inch semi-circle.

Fig. 56. Half-inch Bead and Cove, Coves Completed.

If you are very particular to make the coves in this manner, you will find it much easier to work the piece to a correct shape. If you are careless about making the coves, and especially about cutting the sides square down to the depth of ¼ inch, you will experience much difficulty in turning the beads so that they will be of proper size and shape. You will also be obliged to widen the coves, and in so doing you may have considerable trouble to make them of correct size. This style of turning was used a great deal at one time, as it could be done very rapidly after the turner had learned the necessary motions. These movements would often be learned by continued practice without any attempt to learn the general principles of turning.

LESSON XVII
SPINDLE WITH CONES

Fig. 57 indicates the shape of the spindle with cones, but the size of the various parts may be modi-

Fig. 57. Spindle with Cones.

fied to suit the judgment of the individual student. The general plan of the illustration should be followed. There should be the same number of beads, and they should be similar in size and location. The two halves should be exactly alike, and the parts between the two center beads and the two end beads should be perfect tapers.

ELEMENTARY TURNING

As the two ends are to be exactly the same size, you should measure their diameter with the calipers. Adjust the calipers, as directed in Part 3. Hold them as shown in Fig. 58. Do not force them on to the piece. They should simply touch the two sides so lightly that they will not mar the surface.

At first, you had better stop the lathe while using the calipers. After you have become accustomed to using them, you can do so while the lathe is in motion, if the ends of the calipers are of proper shape.

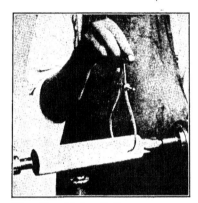

Fig. 58. *Using Calipers.*

Before using the calipers read what is said in Part 3, about shaping the ends for use in wood turning.

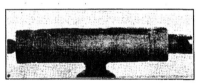

Fig. 59. *Two Beads Turned.*

Turn the entire piece to a cylinder with the roughing gouge, and carefully smooth the surface near each end with the skew chisel. Turn one bead at each end, as shown in Fig. 59.

This piece is to be finished with two coats of shellac, applied with a brush, and, therefore, there must not be any sharp corners which will be rubbed white when sandpapering for the second coat of shellac. Care must be taken that there is no mark left by the point of the obtuse corner of the chisel at the bottom of the grooves or on the sides of the beads. The upper corners of the small spaces at each side of the end beads should be carefully rounded, and the spaces should not be too deep to be smoothed with sandpaper. If there is any space that will gather the finishing material, and is so narrow that the finish cannot be rubbed out, it will injure the appearance of the piece.

Fig. 60. Center Sized.

Keep in mind as the plan is made for the beads and curves, just how the various angles will be sandpapered and finished. After each end has been finished, turn the piece small at the center, as shown in ·Fig. 60. Use the roughing gouge for doing this. The size at the center should be the diameter of the two center beads. Turn the center beads, finishing them smoothly at each side, being careful not to cut the groove between them too deep (Fig. 61). Finish the piece by turning the conical part at each side, finishing with the skew chisel.

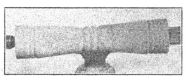

Fig. 61. Center Beads Turned.

ELEMENTARY TURNING 65

The surface of these two cones should be very smooth and straight. The appearance of the finished piece should be as shown in Fig. 55.

LESSON XVIII
SANDPAPERING

For the work in turning, several grades of sandpaper are required. For smoothing the exercises given in Lessons 17 to 25, No. ½ or No. 0 should be used. To produce a very fine finish, use coarse paper at first, and then each finer grade in order, until the required finish has been obtained. No. 00 should pro-

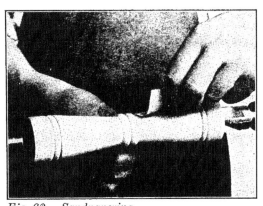

Fig. 62. Sandpapering.

duce a finish fine enough for any school work. The sheet of sandpaper should be torn by using the saw the same as in tearing sandpaper in joinery. Usually, it is best to begin by using pieces but one-eighth of a sheet in size. After some practice in using small pieces, the one-fourth-of-a-sheet pieces may be used. Usually the paper should be folded double. Move the rest out of the way, or remove it entirely before beginning to use the sandpaper.

Hold the paper in both hands, as shown in Fig. 62, always keeping the paper moving from end to end, over the part being smoothed, so as to avoid scratching the surface of the work. The less the paper is moved about, providing it does not scratch the work, the better.

In using a fine grade of sandpaper, you can hold it beneath the work, as shown in Fig. 63. This admits of a better view of the piece. It is not a good plan to use very coarse paper in this manner, as the dust from the wood gathers on the surface of the paper and hinders the flint from cutting. In using the finer grades, the dust is sometimes an advantage, as it causes the sandpaper to cut slower and smoother. In any use of sandpaper, be careful not to throw any more of the dust into the air than is really necessary.

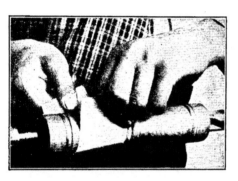

Fig. 63. Sandpapering.

In sandpapering beads, the edge of the folded paper is used, as shown in Fig. 64. As often as the edge becomes worn out, another fold is made. Each time a new fold is made, it should be near the worn one, so that the paper will be used evenly.

In working around beads or curves of any sort, the sandpaper should be given a twist-like motion, in order to preserve the shape of the curve. Sharp edges or deep V-shaped cuts are neither easily sand-papered nor finished, and should, therefore, be avoided as much as possible. The design should be arranged so as to avoid such places.

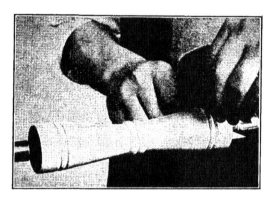

Fig. 64. Sandpapering Beads.

LESSON XIX

SHELLACING

Shellacing of turned pieces may be done with either a brush, or with a polishing pad or a cloth. When the brush is used, the principles involved are the same as those in using the brush on hand-work. One or more coats may be applied, rubbing each with sandpaper or pumice stone.

68   ELEMENTARY TURNING

In applying the shellac with a brush the piece must not be revolved at full speed of the lathe, but simply turned by taking hold of the belt with the hand and pulling, while the brush is held against the work with the other hand, as shown in Fig. 65.

Fig. 65. Shellacing.

In sandpapering the coats of finish, much care must be taken to avoid rubbing entirely through the finish. This is especially liable to occur at the top of beads and at corners. The principle is really the same as in sandpapering hand-work, the apparent

ELEMENTARY TURNING

difference being caused by the speed of the lathe. Often you will have better success if you change the belt to a much slower speed.

There is also a tendency to burn the work, because of the heat resulting from the friction of the paper with the surface revolving so rapidly. The burning of the wood is not so apt to occur, if the finish is ground down with pumice stone and oil. Grinding the finish in this manner is not a very difficult task. Use a cloth or a bit of waste for a grinding pad. Place on the pad a small amount of oil and pumice stone. Ordinary machine oil may be used, but regular rubbing oil is better. Examine your work often, lest you grind off too much of the finish. Use plenty of oil and plenty of pumice stone, for, unless the pad is kept moist and well supplied with pumice stone, it also will burn the work. It should be moved about, similar to sandpaper.

LESSON XX
BEADED SPINDLE

Fig. 66. Beaded Spindle.

After roughing this piece, locate and turn the central bead, as shown in Fig. 67. Do not make

76   ELEMENTARY TURNING

Fig. 75. Marking Spaces.

ends, it should be held as shown in Fig. 75. Never lay it on top of the rest, except after the piece has been smoothed to a cylinder its entire length, as in Figs. 19 and 31, as it is dangerous to do so.

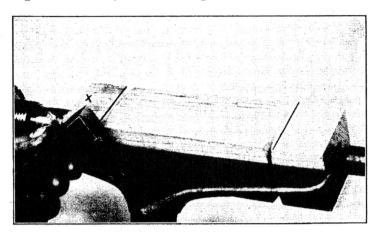

Fig. 76. Cutting in for Square End.

Another way to determine the point for cutting in at the ends is to draw a pencil line across one side of the piece before starting the lathe, or before the

ELEMENTARY TURNING 77

piece has been placed in the lathe. First mark the distances from each end, and then with a try-square or with the rule, used as a straight edge, draw a heavy line entirely across the piece. This line will be visible while the lathe is in motion.

The skew chisel is used to cut the curve at the square part. The first operation is to cut a deep groove, as shown in Fig. 76, with the point of the chisel. To make this V-shaped groove the point of the skew chisel must not be pressed hard against

Fig. 77. Rounding Corners.

the piece, but cut lightly from the two edges of the V until the space is formed.

After the groove has been made, the chisel is reversed, and the corner rounded with the obtuse angle, as in finishing a large bead (Fig. 77). See also Figs. 23, 27, 29, and 33.

Care must be taken in cutting such a place, that nearly all the work shall be done by that part of the edge near the obtuse corner of the chisel. If

72 ELEMENTARY TURNING

rubbing will not be steady or hard enough, but there is danger of rubbing too hard. Only by experience can the amount of pressure be determined. Ridges may be removed sometimes by an increase of pressure, and sometimes the pressure is made so great that the finish is removed or turned black.

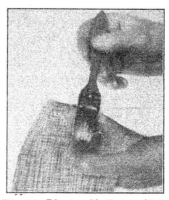

Fig. 70. Placing Shellac on Cloth

Watch the surface very carefully and keep in mind that if too much finish is applied, or if it is not properly rubbed in, it may all be removed by the use of pumice stone and oil, except such places as have been burned. Moistening the cloth with alcohol will sometimes help to remove the ridges

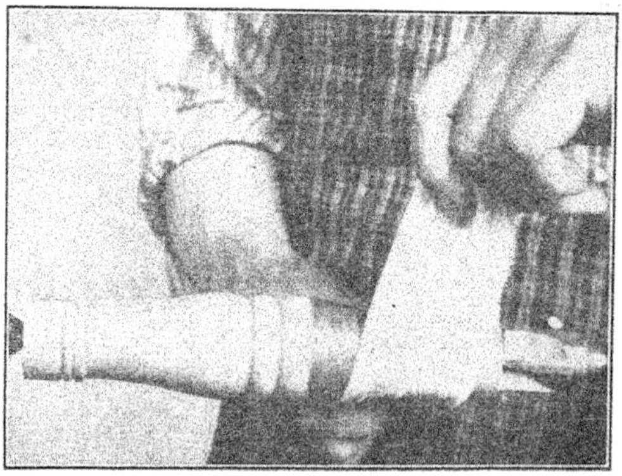

Fig. 71. Polishing with Cloth.

As soon as the cloth begins to stick or pull, a very little oil must be applied to the face of the cloth to keep it from roughing the finish. The oil does the finish no good. In fact, the more oil the poorer the finish, but oil must be used to keep the cloth from sticking.

Some people use the ordinary lubricating oil from the can used about the lathe, but raw linseed oil is used where much polishing is done. For the finest grade of work, rub a little raw linseed oil on the surface before applying the shellac and use more oil during the rubbing if needed. After the polish has been rubbed to a smooth, even gloss, rub with olive oil and then with a clean cloth or the hand, barely moistened with alcohol.

A pad composed of cotton batting or a piece of polishing felt used under the cloth, as shown in Fig. 72, is sometimes of great advantage, but for the present work it is not essential.

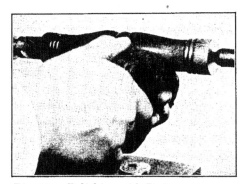

Fig. 72. *Polishing with Pad.*

If the grain of the wood is very open, it is necessary to fill the grain by using a filler, similar to hand finishing of open grained woods. By applying the raw linseed oil before the final sandpapering, the dust from the wood will be moistened and rubbed into the open pores, often making a good filler.

74 ELEMENTARY TURNING

Sometimes all that is required is to coat the piece with shellac, being careful to brush as much as possible into the grain, allowing it to dry thoroughly; after which it should be ground down with pumice stone and finished by rubbing with a cloth, coated with a very little shellac. Do not be satisfied until the finish is even and bright, and the grain is entirely filled.

There are many methods of doing polishing in the lathe, and finishers differ very much in regard to

Fig. 73. Polishing Outfit.

the material to be used, and the method of applying it. It is probably not best to attempt any but the most simple methods, with the simplest of materials at this time.

One item of great importance is to keep the pad or cloth moist. To do this, keep it in a tightly closed dish. A fruit jar or tin can having an airtight cover will be sufficient.

ELEMENTARY TURNING 75

The outfit, shown in Fig. 73, is a very good one. The can is for keeping the cloths and pads which have been filled with shellac. The large bottle is for raw linseed oil. The smaller bottles are for alcohol and olive oil. The square bottle is for rottenstone or pumice stone, and it has a perforated cover. The shellac is taken from the same dish used in joinery.

If the pad or cloth is opened and freshly filled with shellac before it is put away in the can, the shellac will be more evenly distributed when wanted. If a small amount of oil is dropped on to the pad with the shellac, it will generally work better than when applied to the surface. Some finishers mix various gums, also the linseed oil, with the shellac for polishing and for similar work.

LESSON XXII
SQUARE-END SPINDLE

Fig. 74. *Square-end Spindle.*

This lesson introduces the combination of round and square elements on the same piece. In using the rule to measure spaces on a piece having square

76 ELEMENTARY TURNING

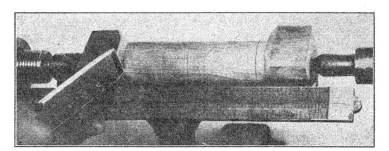

Fig. 75. Marking Spaces.

ends, it should be held as shown in Fig. 75. Never lay it on top of the rest, except after the piece has been smoothed to a cylinder its entire length, as in Figs. 19 and 31, as it is dangerous to do so.

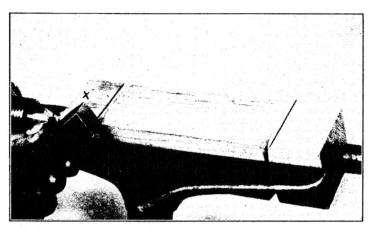

Fig. 76. Cutting in for Square End.

Another way to determine the point for cutting in at the ends is to draw a pencil line across one side of the piece before starting the lathe, or before the

ELEMENTARY TURNING

piece has been placed in the lathe. First mark the distances from each end, and then with a try-square or with the rule, used as a straight edge, draw a heavy line entirely across the piece. This line will be visible while the lathe is in motion.

The skew chisel is used to cut the curve at the square part. The first operation is to cut a deep groove, as shown in Fig. 76, with the point of the chisel. To make this V-shaped groove the point of the skew chisel must not be pressed hard against

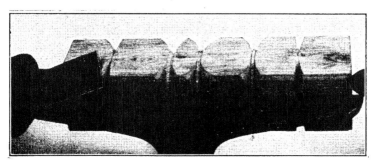

Fig. 77. Rounding Corners.

the piece, but cut lightly from the two edges of the V until the space is formed.

After the groove has been made, the chisel is reversed, and the corner rounded with the obtuse angle, as in finishing a large bead (Fig. 77). See also Figs. 23, 27, 29, and 33.

Care must be taken in cutting such a place, that nearly all the work shall be done by that part of the edge near the obtuse corner of the chisel. If the tool turns much from the vertical position, it will probably catch. When the chisel is properly held, it will not catch, neither will there be much tendency for the tool to jar.

In order to get as much practice as you can before attempting to finish the ends, you can cut several places along the central part of the piece, and round them the same as the ends are to be rounded. Do not make the square part at the ends too short.

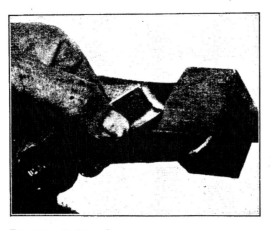

Fig. 78. Rolling Gouge.

After the ends are finished, use the roughing gouge in removing the waste material from the central part of the piece, making it the proper size at the ends for the beads. Roll the gouge, as shown in Figs. 78 and 20, so that it can cut close to the corner without danger of catching.

Next smooth the cylindrical portion with the skew chisel and turn the beads at the ends, as shown in Fig. 79. For this you will probably require the $\frac{1}{4}$-inch skew chisel.

ELEMENTARY TURNING

With the roughing gouge shape the piece as shown in Fig. 80. Instead of using the rule and pencil or rule and skew chisel for marking the position of the center beads, you can set the compasses to the required distance, and by holding them as shown in Fig. 81, make a mark on the piece as it revolves. Do not attempt to make a deep mark with the point of the compasses, but make a light line, and deepen it with the point of the skew chisel, as in Fig. 39.

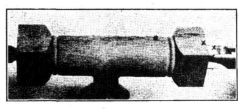

Fig. 79. End Beads Turned.

Fig. 80. Center Sized.

Fig. 81. Spacing with Compasses.

The position of the center bead should be determined by measuring to its sides from each end. The two smaller beads should be measured from the sides of the central bead.

First size the three beads which are at the center, making them square, as shown in Fig. 82; then round these three beads, as shown in Fig. 83.

Complete the piece by turning the long curves. These curves should be roughed to near the finished size with the roughing gouge, and then shaped and smoothed with the 1-inch skew chisel to the form shown in Fig. 74.

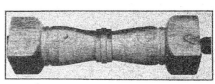

Fig. 82. *Center Beads Roughed to Size.*

Before attempting to do any sandpapering on this piece, remove the rest so that the fingers or the hand cannot be caught between the corners of the revolving piece and the rest.

If the cylindrical portion of the piece is polished the same as the piece in Lesson 21, the flat sides at the ends should be dressed smooth with the plane, and sandpapered and polished by hand after the piece has been turned.

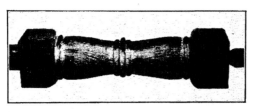

Fig. 83. *Center Beads Turned.*

LESSON XXIII

CURVED SPINDLE

This piece should first be roughed to a cylinder, and then the ends of the curves at the center should

Fig. 84. Curved Spindle.

be finished (Fig. 85). Next the piece should be tapered towards each end, as shown in Fig. 86, using

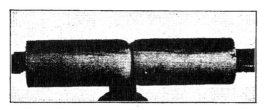

Fig. 85. Curves Started.

the roughing gouge. This determines the diameter of the ovolo at each end.

Turn the ovolo and straight portion at each end (Fig 87), using the chisel for this part of the work. Mark the points A, Fig. 87, with the acute point of the chisel, as in cutting for the cove (Fig.

Fig. 86. Ends Tapered.

39), and then turn this curve with a ½-inch gouge, finishing it as shown in Fig. 88.

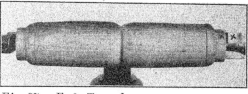

Fig. 87. Ends Turned.

Rough down the long curves with the roughing gouge, and finish with the inch skew chisel to the shape shown in Fig. 84. This piece may be polished by the same method as given in Lesson 21.

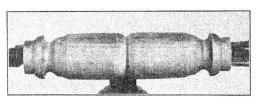

Fig. 88. Cavettos Turned.

LESSON XXIV

TAPERED SPINDLE

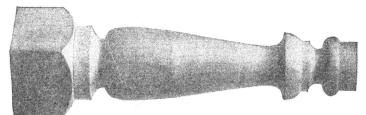

Fig. 89. Tapered Spindle.

This is a form often used in architectural work. The square part of this piece is the lower end or base.

ELEMENTARY TURNING

First turn the curve at the square end and shape the round part as shown in Fig. 90. Smooth the larger end of the cylindrical part and turn the cavetto, which is next to the base (Fig. 91).

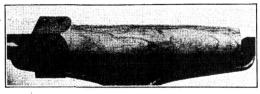

Fig. 90. *Pattern Outlined.*

In finishing this curve, hold the ½-inch gouge as shown in Fig. 92. The gouge is rolled very much to the right, so that it will cut square up to the end of the long curve. The lower end of the long curve is turned with a skew chisel the same as in turning a bead (Fig. 23).

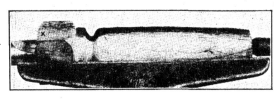

Fig. 91. *Base Completed.*

After the piece has been worked to the shape shown in Fig. 91, a part of the top is turned (Fig. 93). After turning the bead and the straight portion at the extreme top, shape the piece as shown in

Fig. 92. *Gouge on Side.*

84 ELEMENTARY TURNING

Fig. 94. It is very important that this be done properly, so that there will be no need of touching the flat part, F, Fig. 95, after the curve C has been turned.

Work the upper end of the long curve as shown in Fig. 95, and then rough the main part of the curve with the roughing gouge. Finish the long curve with the 1-inch skew chisel to the form shown in Fig. 89. This piece may be sandpapered, but need not be polished.

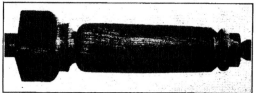
Fig. 93. Top Bead Turned.

Fig. 94. Ovolo Turned.

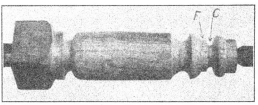
Fig. 95. Cavetto Turned.

LESSON XXV

PORCH SPINDLE

This pattern of spindle is often used on porches, and should be made of pine or other soft wood. The usual lengths of such spindles are 8 inches and 10 inches. The square portion at the bottom is a little longer than at the top. The diameter of the

Fig. 96. Porch Spindle.

bottom bead is nearly as great as can be turned from the piece. The diameter of the top bead is considerably less than the size of the square portion.

Cut in at each end and finish the corners (Figs. 76 and 77).

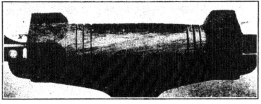

Fig. 97. Beads Spaced.

Rough the central part to as near the desired size as you can with the roughing gouge (Figs. 20 and 78), and smooth each end with the skew chisel. Mark all spaces, as shown in Fig. 97, with the chisel point. Review

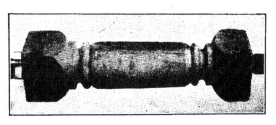

Fig. 98. Beads Turned.

what is said in Lesson 22 about marking spaces on square pieces (Fig. 75).

Turn the cove and bead at the bottom end first, and then turn those at the top, as shown in Fig. 98.

Finish the long curve with the skew chisel to the shape shown in Fig. 96. This spindle should be turned smooth enough for a paint finish without sandpapering.

LESSON XXVI

PLAIN BOX

Select a piece of stock enough longer than the height of the box to allow for waste. For boxes made from $1\frac{3}{4}$-inch squares, there will usually be about $1\frac{1}{2}$ inches of waste. Center the poorest end carefully, and if it is not square with the sides, make it square, either by sawing, planing, or chiselling, before screwing it on to the chuck.

Be careful to bore the hole the correct size so that the screw will hold as much as possible.

Fig. 99. Plain Box.

If you put some tallow or lard into the hole before screwing the piece on to the chuck, you will not only have less difficulty about turning it up tight, but it will hold very much more.

ELEMENTARY TURNING 87

Be sure that it is so tight against the face of the chuck that it will not spring sidewise the least bit.

Set the rest as shown in Fig. 100. The top of the rest should be level with the lathe center, and the end close to the chuck.

Turn the piece to a cylinder, rolling the gouge so it will cut close up to the chuck,

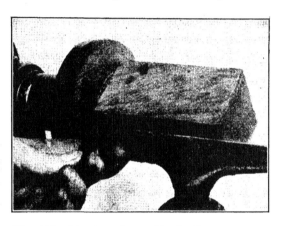

Fig. 100. Blank on Screw Chuck (See Fig. 113).

Fig. 101. Turning Inside of Box Cover.

similar to Figs. 20 and 77. Do not attempt to smooth it, except with the roughing gouge.

Set the rest as shown in Fig. 101, and turn the end for the inside of the cover. To do this, use a gouge at first. The gouge should be held so that it will cut a shaving, but not so that it will catch in the side of the piece. Begin at the center and move it carefully towards the edge nearest to you. As it nears the outer part of the curve, roll it so that the corner will not catch (Figs. 101, 114, and 126). It will cut quite rapidly when properly held.

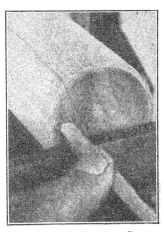

Fig. 102. *Smoothing Inside of Box Cover.*

Finish the curved portion with the round nose scraping tool, as shown in Figs. 102 and 128. The scraping tool is held flat on the rest and quite horizontal. It is moved from the center towards the front side. This tool dulls very rapidly, because the wood passes at a right angle to its edge. The scraping tool is not a cutting tool, and should be used only for finishing. It should be kept sharp, which means that it must be sharpened very often. Read what is said in Part III about sharpening scraping tools.

Cut the square portion, H, Fig. 103, with the acute angle of the skew chisel, holding it as shown in Fig. 104. This part must be made carefully or the cover will not fit properly. If it is not square and

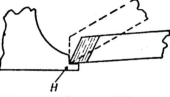

Fig. 103. *Cutting a Rebate.*

sufficiently deep, the cover will not stay in place while you finish the outside.

Fig. 104. Cutting Rebate.

Fig. 103 indicates the position of the skew chisel in cutting the rebate in the cover. The heavy lines indicate the position of the chisel in cutting the outer surface and the dotted lines indicate the position at which the chisel should be held in squaring the bottom of the rebate. Before completing this end, hold the skew chisel against the edge of the cover rim, making it very smooth and square (Fig. 105).

Fig. 105. Squaring End of Cover.

90 ELEMENTARY TURNING

Sandpaper and polish the curved portion, being very careful not to touch the square corners with either sandpaper or shellac. If you do get any shellac into the groove or on to the end, carefully scrape it off after the interior of the cover has been polished.

Start the curve for the top with the skew chisel and cut the cover from the remainder of the piece with the parting tool. This tool will cut either straight into the piece or at an angle. As this cover is not to have a knob, hold the parting tool as shown in Fig. 106.

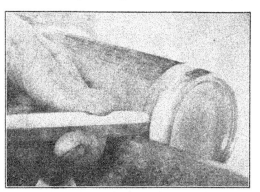

Fig. 106. *Cutting Off Cover.*

Turn a rebate to receive the cover, first cutting with the point of the skew chisel, as shown in Fig. 39, and then lay the chisel flat, as shown in Fig. 107. Be careful to have the cover fit very tight, for it is to be turned on the outside after being put in place. See that the shoulder is square, and that the cover fits so tightly that there is no space at the surface where the two pieces join.

Place the cover in position, and finish the out-

ELEMENTARY TURNING

side, polishing the box on the side and end. Where the cover joins on to the box, the wood may be too thin to work in the ordinary way. If it is, lay the chisel flat and scrape it to size, as in smoothing the end (Fig. 105) and the curve (Fig. 211). It will be necessary to scrape the rounded part of the cover; for, if the chisel is used in the ordinary way, it will shove the cover off.

Fig. 107. *Cutting Rebate.*

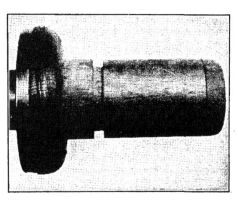

Fig. 108. *Box Ready for Polishing.*

The gouge may be used in finishing the top of the cover, as in turning the bead (Fig. 53) and the cover (Fig. 111). Be sure to have the top of the cover smooth before applying the shellac.

Before polishing the outside of the box, cut a small groove about $\frac{1}{8}$ inch deep at the place where you expect to cut the box off, after it has been finished (Fig. 108).

After polishing all the outside, remove the cover and cut the rebate a little deeper, so that the cover will go on easily. Cut out the inside and polish, being careful to smooth the bottom. Use the gouge and scraping tool for this work the same as in hollowing the inside of the cover (Figs. 101 and 102). If the box is large or deep, the rest may be set as in Fig. 114.

After finishing and polishing the inside, cut the box from the chuck with the parting tool. Write your name on a slip of paper and glue it to the inside of the box.

The stub, remaining on the screw, may be used for a napkin ring, as in Fig. 167, or it may be removed from the screw and placed on an arbor, as in Fig. 174.

LESSON XXVII

BOX WITH KNOB

This box is worked in the same manner as the one not having a knob, except that in cutting the cover off, the parting tool is held at an angle, as shown in Fig. 110. This is to save material for the knob.

Fig. 109. Box with Knob.

ELEMENTARY TURNING

After the cover has been fitted to place, the knob must be turned. To turn the knob, use the gouge as shown in Figs. 53 and 111. This position tends to hold the cover

Fig. 110. *Cover, with Knob Being Cut Off.*

in place because of the pressure of the back of the gouge against the top. Take very light shavings, and work carefully.

After turning the knob finish the outside and the inside the same as the plain box in Lesson 26.

Fig. 111. *Turning a Knob.*

LESSON XXVIII

PLAIN GOBLET

For turning a small goblet the blank is secured to a screw chuck the same as the blank for the box (Lesson 26). Usually the blank is just long enough for the goblet, so that the end of the blank next the chuck will become the bottom of the goblet as in

Fig. 112. Goblet.

Fig. 115. This saves some work and material, and makes the turning easier; because the shorter the piece, the easier it can be turned. If, however, the blank proves to be too long, it can be cut off the same as the box shown in Fig. 108 or the goblet in Fig. 122.

Be very careful to have the piece screwed on so tightly that it will not spring away from the facing in the least. Set the tee rest as in Fig. 113, also see Fig. 100, and rough the blank down to a cylinder. Stop the lathe and examine the piece carefully. The blank may have been resting at the corners, and now that they have been cut away, the piece may require tightening. Set the rest as in working the inside of the box cover (Fig. 101) and shape the inside of the

Fig. 113. Rounding a Blank.

ELEMENTARY TURNING 95

bowl, also square the end as shown in Fig. 105. Sometimes the tools will cut better, if the end of the rest is set into the bowl, as shown in Fig. 114. The inside should be turned very smooth, using first the gouge and then the round nosed scraping tool. Be sure that the scraping tool is very sharp.

You will need to use quite coarse sandpaper at first. No. 1½ will probably be the best grade, unless the goblet is very large and of a coarse wood. If it is, use a little of No. 2 sandpaper. Hold the paper so it will not spoil the edge or rim of the bowl. To avoid this, you may need to tear

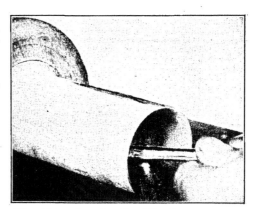

Fig. 114. Rest Inside of Bowl.

the paper to ⅛-size or perhaps even smaller. Be sure to smooth the bottom end of the bowl, and also to sandpaper down any ridges on the inside.

Finish the inside entirely, including the polishing, for it is not best to attempt to polish the inside after turning the outside of the bowl. Each time a part is polished, it should be so well done that it will not be necessary to touch it again, for after the outside of the bowl is finished it is too thin to be polished on the inside; and after the stem has been turned the bowl will probably revolve a little out of true, and the outside cannot then be polished.

The next part of the work is to turn and polish the outside of the bowl, polishing it to the small shoulder, S, Fig. 115. The reason for working to the point, S, is that the square corner at this point is a good place at which to join the two parts of the finish.

After the outside of the bowl has been polished, turn the base and stem. Be very careful not to allow your tool to slip and spoil the base. Smooth the outer edge of the base with the skew chisel.

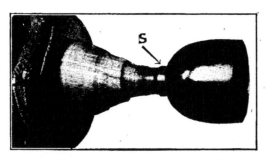

Fig. 115. Goblet Bowl Polished.

Make several marks with the point of the chisel to assist in starting the gouge, which should be held as shown in Fig. 116. Be sure to have the edge of the base so thick that the pressure of the back of the gouge will not break the edge.

Turn the large curve with the gouge in the same manner as you turned the 1-inch cove. Use such tools in turning the stem as the pattern requires. Polish the base and the stem, and then remove the goblet from the chuck.

ELEMENTARY TURNING 97

After the goblet has been removed from the chuck, the bottom may be rubbed on a sheet of

Fig. 116. *Turning Base of Goblet.*

sandpaper, laid on the bench or a flat board. Do not rub the bottom surface much, or the edge will be injured.

LESSON XXIX

GOBLET WITH RINGS

If rings are to be turned about the stem of the goblet, they are worked from the material that is ordinarily cut away. In turning the outside of the bowl leave as much material for the rings as you can. Compare Fig. 115 with Fig. 118.

Fig.117. *Goblet with Rings.*

98 ELEMENTARY TURNING

To turn the rings, first turn beads, and then use the ring tools on each side, as shown in Figs. 198 and 295. Gradually work around each ring, until it is nearly cut from the piece. If there is but one ring, it may be worked entirely with the skew chisel, as shown in Fig. 164. If there

Fig. 118. Goblet Bowl Polished.

are several rings the ring tools will be required. Sandpaper and polish the rings carefully, as shown in Fig. 119 and then, by using the ring tools, cut the rings entirely free.

After the first set of rings has been cut loose, another set may be made, as shown in Fig. 120; and after these have been finished, yet another set may be made.

Fig. 119. Goblet Rings Polished.

By making rings small and close together, a large number may be cut on one goblet. Before

ELEMENTARY TURNING

cutting off the second set of rings, turn as much of the base and stem as you can in order to make room for the rings when loosened.

Figs. 120 and 121 show how to hold the rings with one hand, while turning the base and stem with the other hand. Fig. 121 also shows the groove made in starting to cut the goblet from the chuck

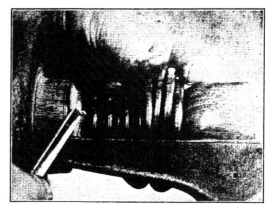

Fig. 120. Turning Base of Goblet (See Fig. 116).

The blank for this goblet was too long for the size of the top, therefore it was necessary to cut it off, as shown in Fig. 122.

In using the parting tool in such a place, start the cut with the point of the skew chisel to avoid roughing the

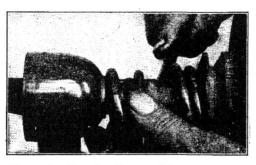

Fig. 121 Turning Between Rings.

edge. Hold the chisel at such an angle as will cause the goblet to rest on the outer edge of the base. By cutting a wide space, the bottom of the base may be sandpapered to near the center before cutting it off.

In cutting off a piece of this shape, it is better to cut well in from the surface with the acute point of the skew chisel, leaving only a small part to be

Fig. 122. Cutting Goblet from Chuck.

cut with the parting tool. If there is plenty of room, the piece may be cut entirely off with the skew chisel. This will make a better finished bottom. In either case it may be sandpapered after being removed from the lathe

After the goblet has been removed from the chuck, the inside of the rings should be smoothed with the knife and sandpaper, and then polished by hand.

LESSON XXX
A ROSETTE

Rosettes are made in many designs. They are usually held on a screw chuck while being turned. Fig. 124 illustrates the blank in place, and the turning gouge cutting the outer edge. If the blanks have been carefully sawed to shape on a band saw, the outer edge will not require much turning. If they have been cut to shape by simply sawing the corners off

Fig. 123. Rosette.

with a hand saw or a back saw, as in Fig. 259, there will be danger of breaking the gouge, if the piece is of hard wood, unless you work very carefully. If the edge is very rough or the wood very hard, hold the gouge nearer on a level so that it will not cut too deeply.

Fig. 124. Turning Edge of Rosette.

Turn from each edge towards the center, rolling the gouge, as shown in Figs. 124 and 125.

Whether the face surface of the rosette should be turned with

Fig. 125. *Turning Edge of Rosette.*

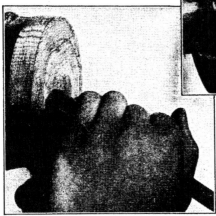

Fig. 126. *Turning Face of Rosette.*

the roughing gouge or a turning gouge will depend upon the pattern. For this design you should use a roughing gouge for the general outline, and then shape the parts with the turning gouge, finishing with the round end scraping tool and firmer chisel. Fig. 271 shows how the roughing gouge is held.

Fig. 127. *Scraping with Chisel.*

Fig. 126 shows how the turning gouge is held to make it cut instead of scrape. Before attempting to use the turning gouge in this manner, refer to Figs. 231, 232, and 233.

Scraping tools must be used to finish the surfaces, both at the edge and on the face.

Figs. 127 and 128 show how these tools are held. Figs. 102, 171, and 211 show other positions of scraping tools in use.

Fig. 128. Scraping with Round Nosed Tool.

SUPPLEMENTARY EXERCISES
PART II
INTRODUCTION

The thirty lessons cover substantially all the ordinary uses of wood turning tools. More difficult problems are easily suggested, yet, for the time usually devoted to turning, it is not advisable to undertake the more difficult problems.

For those who are more apt in this line of work, or who wish to devote more than the ordinary time to this subject, additional exercises are given. There are also a variety of designs which may suggest other problems and combinations requiring no additional directions.

It is better to execute the exercises in this part in their given order, yet, by a thorough use of cross references, a pupil who has completed Part 1 should be able to make any of these articles properly. Until all the work in Part 1 has been completed, no attempt should be made to do any of the work in Part 2.

NUMBER I
TOOL HANDLE

Tool handles may be made from pieces which are too small for regular exercises, or out of pieces which have been accepted as exercises and then discarded.

Usually the piece is turned with the roughing gouge to the general outline, as shown in Fig. 130. With the point of the skew chisel (Fig. 39) the length of the space for the ferrule is marked. The end is then turned to a size that will allow of the ferrule being driven to place.

Fig. 129. *Tool Handle.*

Turn the conical part, which is next to the ferrule, using the roughing gouge and the skew chisel. With the turning gouge (Fig. 47) turn the small curve. Work the main part of the handle to size with the roughing gouge (Fig. 13). Finish the handle to the shape shown in Fig. 131, using the skew chisel the same as in finishing the body of Fig. 89.

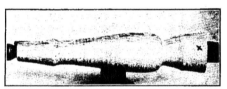

Fig. 130. *Tool Handle Roughed.*

After the handle has been sandpapered and polished, as shown in Fig 131, carefully cut the stub end off with the skew chisel while the piece is in the lathe. Do this so that there will be no roughness on the end of

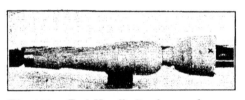

Fig. 131. *Tool Handle Sandpapered.*

the handle. There should be no stub end at the dead-center bearing.

Another method of making a handle is to work several places to near the finished size, as indicated by Fig. 132. The measurement should be made as given in drawing, Fig. 133. The parting tool is used for the cutting or scraping, and the calipers are used for measuring the diameters.

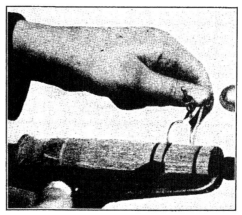

Fig. 132. *Tool Handle Sized.*

The piece is then worked to size by carefully cutting with the roughing gouge to the bottom of the grooves made by the parting

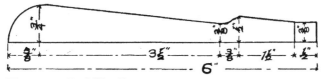

Fig. 133. *Tool Handle.*

tool. The handle is then finished in the same manner as first described.

Fig. 129 illustrates an ordinary file handle. Handles are made of all sorts of shapes and sizes, and of many kinds of wood. The cheaper grades of file handles are of soft wood. Chisel handles should always be made of hard wood. Apple wood is often used for firmer chisel handles. Socket chisel handles for heavy work are often made from hickory. Sometimes they are fitted with an iron ferrule at the top end, similar to the ring on the mallet (Fig. 150).

Fig. 134. *Leather Topped Handle.*

Handles which are to be struck with a mallet may be flat at the top end, and have two or three thicknesses of leather glued or nailed to the end, so as to hinder the mallet from splitting the handle. Fig. 134 shows a handle of this kind for a tanged firmer chisel. Handles for socket chisels may also have leather tops.

Fig. 135. *Socket Chisel Handle.*

Fig. 135 illustrates a handle of fancy pattern for a socket chisel. These handles are not to be struck with a mallet.

ELEMENTARY TURNING

NUMBER II

GAVEL

The essential features of a gavel are that the ends of the head should be rounding in shape, and that it be made of wood from a hard, sonorous variety, and of a pleasing design.

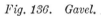

Fig. 136. Gavel.

Turn the head of the gavel first, being particular to finish the wood so that it will take a very high polish. Be sure to select stock long enough to allow for waste at each end, as shown in Fig. 137. Usually, there should be a longer stub of waste at the spur center than at the dead-center.

Fig. 137. Gavel Roughed to Shape.

Outline the pattern, as shown in Fig. 138. Finish the central portion of the pattern first, and then work toward the ends, finishing the rounded ends last. These will require very careful scraping in order to finish properly. Hold the chisel while scraping as shown

110 ELEMENTARY TURNING

in Figs. 105, 127, and 211. After the piece has been polished, it will appear as in Fig. 139.

In cutting the stub end off, leave enough material to sandpaper thoroughly so that each end will be free from any roughness or marks caused by the tools. To sandpaper the ends, lay a piece of sandpaper on the bench, and rub the gavel on the paper. Use coarse sandpaper at first in order to cut the end down to an even, rounding surface. After removing the rougher places, finish the work by holding the sandpaper in the palm of your hand. Finish with paper so fine that the ends will take a polish equal to the other parts of the gavel.

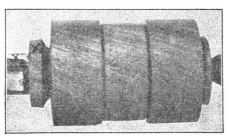

Fig. 138. Pattern Outlined.

Examine the head carefully and plan to have the handle located

Fig. 139. Head Finished in Lathe.

so as to give the best effect. Place the head in the vise with a block at each end, as shown in Fig. 140. Bore the hole nearly through, being careful to make it straight and at right angles to the surface.

ELEMENTARY TURNING 111

Select a piece for the handle, and turn it as indicated in Fig. 141, having the larger end near the live-center. First round the piece the entire length, and then fit the end next to the dead-center into the hole bored in the head of the gavel. Determine the length and turn the large end. Finish the central portion (Fig. 142), and then carefully polish all, except the part which enters the head. Glue the handle to place.

Fig. 140. Gavel Head in Vise.

Gavels vary greatly in size. No. 136 is $3\frac{1}{2}$ inches long and $2\frac{3}{8}$ inches in diameter. The handle is $9\frac{1}{2}$ inches long and $1\frac{3}{16}$ inches

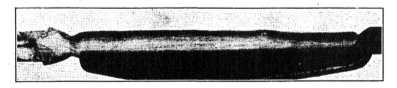

Fig. 141. Handle Roughed Out.

in diameter at the large end. The hole in the head for the handle is $\frac{1}{2}$ inch in diameter.

112 ELEMENTARY TURNING

No. 144 is 2⅞ inches long and 1⅞ inches in diameter. The handle is 9 inches long and ¾ inches in diameter at the large end. The hole in the head is 7-16 inch.

Fig. 142. *Handle Polished.*

NUMBER III

GAVEL PATTERNS

These patterns represent a variety of possible forms for gavels. Study them carefully, and then work out a design of your own. Follow the plan outlined in making number 136 by first planning the length, then working the center, and lastly turning the ends.

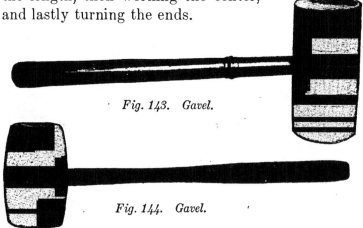

Fig. 143. *Gavel.*

Fig. 144. *Gavel.*

ELEMENTARY TURNING 113

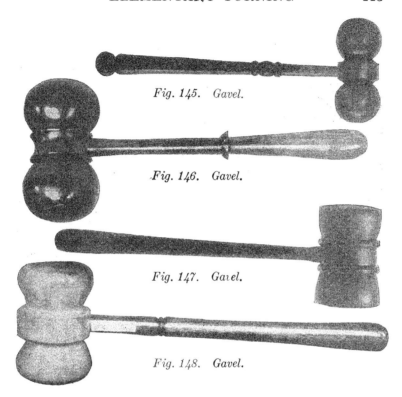

Fig. 145. Gavel.

Fig. 146. Gavel.

Fig. 147. Gavel.

Fig. 148. Gavel.

NUMBER IV
CARPENTER'S MALLET

The carpenter's mallet is worked in the same manner as the gavel (Fig. 136). Such mallets vary greatly in size. The one shown in Fig. 149 is 5 inches long, and $2\frac{3}{4}$ inches in diameter. The handle is 10 inches long, including the 2 inches in the head, It is $1\frac{1}{8}$ inches in diameter at the large end and $\frac{5}{8}$ inches in diameter in the head.

Mallets for such use should be plain and larger at the center than at the ends. There should be no deep markings or grooves in the head near either end, for such a breaking

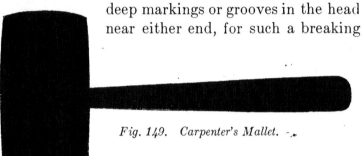

Fig. 149. Carpenter's Mallet.

of the surface will cause the splitting of the mallet.

The two dark bands in Fig. 149 are not deep-cut beads. They were made by cutting very small grooves at each edge of each band and holding the end of a stick against the surface while the lathe was in motion, until the surface of the wood was darkened.

The handle should not have any ring, shoulder or other break in the surface at the end near the head, for such a design will cause

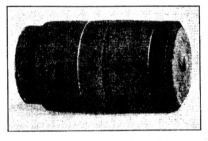

Fig. 150. Mallet with Iron Rings.

the strain when in use, to concentrate at one point which will soon cause the handle to break.

For heavy work, mallets have iron rings to keep the wood from splitting. The ends should be cut

ELEMENTARY TURNING 115

down to receive the rings. The wood should extend out beyond the ring so that as the mallet is used it will batter over the ring and hinder it from coming off, otherwise the ring will jar off in use. The rings should be heated and shrunk on. Fig. 150 shows the mallet head with the ring on one end and the space ready for the ring at the other end.

Carpenter's mallets are made from box-wood, hickory, maple and similar woods. Sometimes a very tough knot or knurl is used for a mallet.

NUMBER V
CARVER'S MALLET

Figs. 151 and 152 show two designs for carver's mallets. The usual sizes are indicated by the drawing, Fig. 153. No special directions are required for making them. The bottom or large end should be straight across so that they will stand on end when not in use.

Fig. 151. Carver's Mallet.

Fig. 152. Carver's Mallet.

ELEMENTARY TURNING

Fig. 153. *Carver's Mallet.*

NUMBER VI

MOLDER'S RAMMER

Fig. 154. *Molder's Rammer.*

Fig. 154 illustrates a molder's rammer. This should be made from hard wood, maple being one of the best for this purpose. Cheaper woods, such as beech, may be used. The sizes given in the drawing (Fig. 155) are for a rammer, suitable for use by pupils in the high school.

First, turn the piece to a cylinder; next, cut away the central portion, making the handle. Finish the ends, round all the corners, sandpaper and

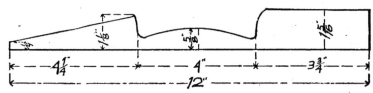

Fig. 155. *Molder's Rammer.*

ELEMENTARY TURNING

oil the entire surface. Remove the piece from the lathe, and then finish the ends (Fig. 156).

Lay out the long end, and saw and plane it to the finished shape, as shown in Fig. 154.

Fig. 156. Turning for Molder's Rammer.

NUMBER VII

DARNING BALL AND DARNING HEMISPHERE

The darning ball (Fig. 157) and the darning hemisphere (Fig. 159) should be finely finished, especially on the large hemispherical ends. The sizes may vary. Those given in

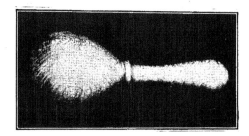

Fig. 157. Darning Ball.

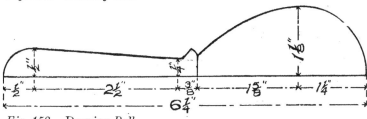

Fig. 158. Darning Ball.

the drawings (Figs. 158 and 160) are of the average size. The wood used should be of a close, hard grain. Maple, cherry, apple, etc., are suitable.

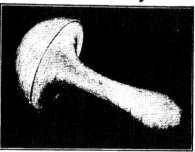

Fig. 159. Darning Hemisphere.

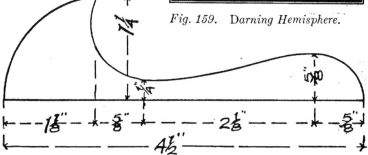

Fig. 160. Darning Hemisphere.

NUMBER VIII

GLOVE MENDER

Fig. 161. Glove Mender.

This glove mender should be made from hard wood, and about 4½ inches long. The ends should be of a size to fit the glove fingers. It should be very carefully smoothed.

NUMBER IX

PLAIN RING

This exercise is given to show a method of making a ring without the use of special tools. Fig. 163 shows a piece which was cut from the bottom of a box.

Fig. 162. *Plain Ring.*

A ⅝-inch hole was bored through the center of the piece, and it was forced on to an arbor Read what is said about arbors in Part III.

Turn the outside of the ring in the same manner as an ordinary bead is turned (Figs. 23 and 33). With the skew chisel held as shown in Fig. 164

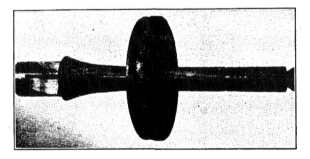

Fig. 163. *Ring Blank on Arbor for Turning.*

scrape around towards the inside of the ring from both the right and the left sides.

In this manner shape the ring, until it is nearly severed from the waste material. Polish the ring carefully, reaching as far inside as you can.

Fig. 164. Turning the Ring.

After the ring has been polished, cut it entirely free, holding the chisel the same as in Fig. 164. Rings may be made on the stems of goblets in this manner. Smooth and polish the inside of the ring by hand, after removing it from the lathe.

To finish the ring on the inside, it may be placed in a chuck, as shown in Fig. 165. After one side has been polished, reverse the ring and polish the other side.

Instead of using the "cut and try" method of making the hole in the chuck, you may set the inside

Fig. 165. Ring in Chuck.

ELEMENTARY TURNING 121

calipers and hold them as shown in Fig. 263. Do not attempt to touch them to the chuck while it is in motion.

NUMBER X

NAPKIN RING, FIRST METHOD

Napkin rings may be made of many shapes and sizes. The scraps of wood left from regular exercises, boxes, etc., can be used for making them. The larger sizes are made about 2 inches in diameter.

Fig. 166.
Napkin Ring.

Fig. 167. Ring Turned on Screw Chuck.

Usually the grain of wood should be parallel with the axis of the ring, but sometimes it may be at right angles to the axis.

Only pieces quite free from defects should be used, for

122 ELEMENTARY TURNING

when the ring has been turned to shape, it is so thin that a small check is quite likely to cause it to break.

The outside, inside, and one end may be turned on the screw chuck, as shown in Fig. 167. This is the piece left after turning the box (Figs. 99 and 108).

After completing the ring as shown in Fig. 167, fasten a piece of pine to an iron face-plate (Fig. 278), using at least four screws. Be careful to locate the screws so that they will not be in the way of the tools in making the place for the ring.

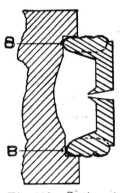

Fig. 168. Section of Napkin Ring in Cup Chuck.

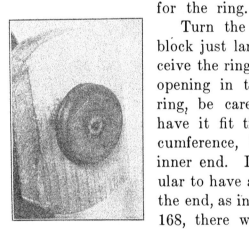

Fig. 169. Napkin Ring in Cup Chuck.

Turn the opening in this block just large enough to receive the ring. In making the opening in the chuck for the ring, be careful to not only have it fit tightly at the circumference, but also at the inner end. If you are particular to have a good bearing at the end, as indicated at B, Fig. 168, there will be much less difficulty about the piece running true. It is not necessary that the hole in the face-plate be deep. It is

ELEMENTARY TURNING

sufficiently deep if it admits the ring beyond the center of the first bead.

If both ends are to be chucked, be careful to fit the hole to the smaller end first. See that the ring is held firmly and that it revolves true.

Fig. 169 shows the ring in the chuck. Set the rest as in Fig. 170, and then bore out the inside with the gouge, finishing with the skew chisel, held as shown in Fig. 171. The skew chisel may be used as a scraping tool, on the inside of the ring, if the rest is placed so that the edge of the chisel is a little above the center of the ring. If it is below the center, it is likely to enter too deeply into the wood.

Fig. 170. *Boring Napkin Ring.*

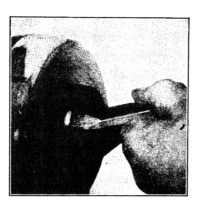

Fig. 171. *Skew Chisel Smoothing Ring.*

Fig. 172. *Polishing Ring on Arbor.*

124 ELEMENTARY TURNING

Finish this end by sandpapering and polishing it completely. Reverse the ring and refinish the other end.

If in chucking, the finish on the outside should be injured, the ring may be placed on an arbor, as shown in Fig. 172, and refinished.

NUMBER XI

NAPKIN RING, SECOND METHOD

Instead of screwing a blank on to a screw chuck, it may

Fig. 173. Napkin Ring.   Fig. 174. Napkin Ring Blank.

Fig. 175. Outside of Napkin Ring Finished.

be placed on an arbor the same as the plain ring, Fig. 163. It is not necessary that the blank be cylindrical. Fig. 174 shows a blank in place. The tools must cut lightly to avoid causing the

ELEMENTARY TURNING

arbor to turn in the hole. Turn the piece to a cylinder, cut the ends square, and then shape the pattern.

Fig. 175 shows the outside of a ring finished, and the piece ready to be removed from the arbor and placed in a cup chuck. It is then finished the same as Fig. 169.

Fig. 177. Beaded Napkin Ring.

Fig. 176. Plain Napkin Ring.

Figs. 176 and 177 show other designs for napkin rings.

NUMBER XII

VISE HANDLE

Select a piece of straight grained hardwood, 12½ inches long, and turn the end, as shown in Fig. 179. Move the rest and turn the other end, as shown in

Fig. 178. Vise Handle.

Fig. 180. Bore a hole in a block and force it on to the end, as shown in Fig. 181. Turn the block to shape, completing the handle. Sandpaper all, and

Fig. 179. One End of Vise Handle Finished.

cut the handle out of the lathe, the same as the tool handle (Fig. 129). Fig. 178 shows the completed handle.

After the handle has been placed in the iron at the end of the vise screw, glue the wooden knob

Fig. 180. Handle Ready for Knob Blank.

to place. Do not use any brad in the knob. In order to make the handle from smaller stock, dealers sometimes turn both balls separate from the bar.

Fig. 181. Knob Blank in Place.

NUMBER XIII

WOODEN SCREW

For wooden hand screws, two kinds of screws are required. The back screw, shown in Fig. 182, and the shoulder screw having a square shoulder next the handle, as shown in Fig. 183.

Fig. 182. Back Screw for Wooden Hand Screw.

About the only wood suitable for these screws is straight-grained hickory.

Finish the handle first. Re-set the tee rest and finish the other end the same as in making the vise handle (Figs. 179 and 180).

Fig. 183. Turning Wooden Blank for Screw.

The part on which the thread is to be cut must be smooth and of correct size. By holding the hand,

128 ELEMENTARY TURNING

as in Fig. 183, the piece can be smoothed with the skew chisel. Test the piece carefully with the calipers. It should be of the same size as the smooth

Fig. 184. Cutting the Threads for a Wooden Screw.

part of the hole in the screw box, which is to be used in cutting the thread.

Do not use any sandpaper on the part of the piece which is to be threaded. Start the thread

ELEMENTARY TURNING 129

by hand, and then place the piece in the lathe. Put the belt on a slow speed. Remove the rest, so that if the screw box becomes stuck, it will not be injured by striking the rest. Move the shifter just enough to cause the piece to revolve slowly (Fig. 184). Some tallow thoroughly rubbed on the wood will cause the screw box to cut a smoother thread.

NUMBER XIV

LARGE BOX

This box differs from those shown in Figs. 99 and 109, not only in the shape of outline, but also in the fitting of the cover. The cover is turned in a manner

Fig. 186. *Inside and Bead of Cover Polished.*

Fig. 185. *Box.*

similar to the one shown in Lesson 27, the difference being that the inside is a simple curve, and the shoulder for fitting against the top of the box is on the outside, as shown in Fig. 186.

130 ELEMENTARY TURNING

The large bead at the top end of the box and the rounded edge of the cover are to obscure this joint.

The inside of this cover and also the bead should be polished before the cover is cut from the box. The end of the box should be polished before the cover is put in place, because of the difficulty in polishing the small groove between the end of the box and the cover, after the cover is on the box.

This is a large box, being made of 4-inch stock, and you should do most of the work with the skew chisels and gouges rather

Fig. 187. *Turning the Bead on the Cover.*

ELEMENTARY TURNING

than with scraping tools. Fig. 187 shows a good position for using the skew chisel in turning the large end.

Fig. 188 shows the cover in place, the outside polished, and the groove at the base where the box

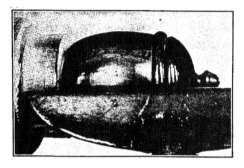

Fig. 188. Box and Cover Polished.

Fig. 189. Working Out the Inside of a Box.

will be cut off. Fig. 189 shows the position of the rest and the gouge in boring the inside. Notice that the gouge is rolled towards the left so that it will not dig into the work.

In removing the waste from the interior of a box of this size, usually the ½-inch gouge can be used. When this gouge is used, the rest is generally set square across, as shown in Fig. 189, yet it may be set as in turning the light goblet (Fig. 114).

Unless the gouge is so light that it springs, the rest had better be set square across.

This box is large enough to receive a very nice polish on both inside and outside.

NUMBER XV

BOXES

There is opportunity for a very great diversity of form and size in the designs of boxes. Some may have their covers tightly fitted as shown in Figs.

Fig. 190. Box. *Fig. 191.*

ELEMENTARY TURNING 133

99 and 109; others may have covers as in Fig. 185 or 191. The body of the box may be straight, curved, or ornamental. After examining these designs, make a design of your own.

Fig. 192. *Fig. 193.*

NUMBER XVI

CANDLESTICKS

Although candlesticks of wood are not very useful, yet they are excellent exercises in turning. They may be supplied with a metallic top.

Usually, in making the candlestick it is more convenient to turn the parts separately, finishing them completely, including polishing, before putting them together.

The base may be fastened to an iron face-plate and finished, as shown in Fig. 195. The hole is bored the same as in making the box, Lesson 26. If you wish to finish the under side of the base, it may be placed in a wooden chuck, the same as the ring (Fig. 165); or the napkin ring (Fig. 169); or the pin tray (Fig. 227).

Care must be taken to locate the screws so that they will enter the thick portion of the base, or they will interfere with the turning.

The stem is turned on the center the same as the exercises in Part I. Fig. 196 shows the piece outlined, and Fig. 197 shows it finished.

The joint at the base may be hidden by a bead as in

Fig. 194. Candlestick with Handle.

Fig. 194, or it may be shaped as shown in Fig. 199 or 203. The hole at the top for the candle is usually bored with a Forstner bit after all other work on the candlestick has been completed. The hole may be started while the piece is in the lathe. For this purpose use either a small turning gouge or a turning chisel. Sometimes

Fig. 195. Base for a Candlestick.

a ring for a handle is placed on the side, as shown in Fig. 194. The ring is turned and finished on an arbor, as shown in Fig. 198. This ring is too small to be turned easily with the skew chisel, therefore, the special ring tools are used, as shown in Figs. 198 and 295. These tools may be of various forms and sizes, as described in Part 3.

Fig. 196. Candlestick Stem Outlined.

Fig. 197. Candlestick Stem Finished.

After the ring has been polished as much as possible on the arbor, cut it entirely loose, and finish it by hand or in a chuck, as shown in Fig. 165. Fit with a knife where it joins the base and the stem. Scrape the finish off wherever the parts join so that glue will hold properly, and glue it to place.

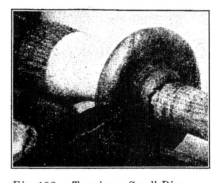

Fig. 198. Turning a Small Ring.

In turning candlesticks, similar to Fig. 199, the base is turned on a chuck the same as shown in Fig. 195. The stem is roughed

136 ELEMENTARY TURNING

out between centers, and the joint carefully made by holding the tools as shown in Figs. 206 and 207. The stem is then glued to place, and turned on the face-plate. The joint must be thoroughly sized with glue before putting together.

Fig. 199. *A Low Candlestick.*

NUMBER XVII
DESIGNS FOR CANDLESTICKS

Fig. 200.   Fig. 201.   Fig. 202.   Fig. 203.

By studying the four designs on the preceding page, and Nos. 194 and 199, you ought to be able to make an original design.

NUMBER XVIII

HAT RESTS

The hat rest, shown in Fig. 204, may be made by turning the base on a face-plate, the same as the base of the candlestick, Fig. 195.

The top should be turned on an arbor, similar to the napkin ring, Fig. 174. The stem may be turned on the centers the same as the first exercises or the candlestick stem (Figs. 196 and 197), and the ends fitted to the holes in the base and top.

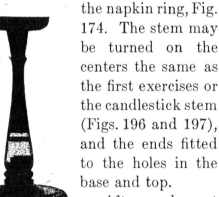

After each part is completed, all should be glued to-

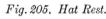

Fig. 204. Hat Rest.   Fig. 205. Hat Rest.

gether. If the hat rest is to be made in this manner, there may be beads or some similar curves at the joints so that any variations will not be noticeable, as in the candlestick, Fig. 194.

Another method, and one which may be used for such designs as Fig. 205, is to rough out all the parts and fit them together. The piece for the top should be roughed to shape on an arbor. In squaring down the end, the skew chisel may be held as shown in Fig. 206

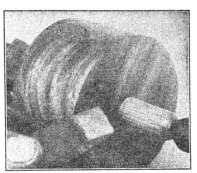
Fig. 206. Jointing Side of Blank.

To finish the surface, the skew chisel should be held as in Fig. 207, and a very light cut should be taken.

Regular scraping tools, if they are at hand, should be used for facing these pieces.

The stem should be roughed to the shape shown in Fig. 208.

The ends should be carefully formed. The surface which forms the joints must be very well turned, or the joint will show badly after the parts are polished.

Fig. 207. Jointing Side of Blank.

Fig. 208. Stem for Hat Rest.

The base should be secured to a face-plate, the

ELEMENTARY TURNING

same as the base in Fig. 195 or Fig. 258. The edge must first be turned, using the gouge as in Fig. 125 or 126. No attempt should be made to scrape these surfaces, for the gouge will again be used on them, after the parts have been glued together.

The face of the base should be turned with the roughing gouge, as in Fig. 271, and then a hole bored for the pin which is to extend entirely through the piece. You may find it of advantage to use longer screws in fastening the blank to the face-plate, so that

Fig. 209. Base Roughed Out.

you can block it away from the face-plate by placing strips about $\frac{1}{4}$ inch or $\frac{3}{8}$ inch thick between the base block and the iron.

After the hole has been finished, turn the base to the form shown in Fig. 209.

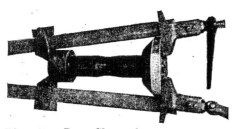

Fig. 210. Parts Clamped.

Glue the three pieces together, clamping them with strong clamps, as shown in Fig. 210.

In making such joints in pieces to be turned, the end grain and the side grain should be thoroughly sized with glue before gluing together, so that in turning the grain will not be torn or broken at the joint. If the sizing is properly done, the joint can be turned after drying as perfectly as if it were one piece.

You must plan to preserve the centers on the stem, so that they can be used after gluing. The spur center will not hold so strongly as the screws in the face-plate, therefore you must be very careful in turning the base and top after the parts have been glued.

Fig. 211. Hat Rest Outlined.

The first thing to do after the parts have been joined, is to go over the entire pattern, turning each part to nearly the finished size. Fig. 211 shows the piece nearly to size, and also shows how the chisel is held to scrape the face of a curve.

To form the long curve, work carefully from each end. The top and bottom parts should be nearly finished before turning the stem. Fig. 205 shows the finished hat rest.

NUMBER XIX
COMBINING WOODS

To combine woods for ornamental turning does not require a great deal of skill. The object should be to combine them so that the effect will be pleasing rather than novel.

There are two ways usually employed in preparing such work for turning. The one more often used, and probably the better, is to glue together thin boards of two or more varieties of wood, making a block sufficiently large to allow of ripping across the glue joints after the boards are all in place. This will result in a block having a cross section as indicated in Fig. 212

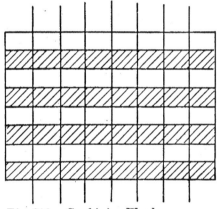

Fig. 212. Combining Woods.

The block is then ripped across the glue joints, as indicated by the vertical lines, making a number of boards equal in thickness, after being smoothed, to the thickness of those first used. These pieces are reversed and glued together, making a block. The end of this block will be a series of exact squares, as shown in Fig. 213, if the work has been properly done.

Unless the pieces are all made of exactly the same thickness, the squares will not meet exactly, and the turned piece will not look well when finished.

The joints must be very perfect, or they may open after the piece has been finished. This blank is then turned in the lathe to some shape that will show the combined woods to the best advantage. Fig. 214 is an illustration of a goblet made in this manner.

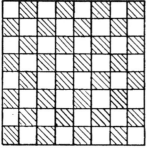

Fig. 213. Combining Woods.

Another method is to select a piece for the central portion and glue to it such shaped pieces as are desired. The pieces may be of almost any shape or size, but should be in pairs or groups, so that the object, after being turned, will show a well defined pattern. In all this work, be very careful to have the pieces of exact size and their surfaces in perfect contact. In the box, Fig. 215, the small pieces were glued around a central piece.

Fig. 214. Fancy Goblet.

Fig. 215. Fancy Box.

NUMBER XX

DESIGNS FOR GOBLETS

Although the wooden goblet is more ornamental than useful, yet as a turning exercise, it is quite valuable. The following designs will suggest many others.

Fig. 216. Goblet.

Fig. 217. Goblet.

Fig. 218. Goblet.

Fig. 219. Goblet.

NUMBER XXI

KNIFE AND FORK REST

This article is a very good exercise to illustrate the use of a templet. Read what is said in Part 2 in regard to the making of templets, and make one for the piece you are about to turn.

Fig. 220. Knife and Fork Rest.

If you have any doubt in regard to the size which you will be able to make from the blank, turn it to a cylinder and space it, as shown in Fig. 221, and then calculate the sizes for the templet.

Rough the piece so it will nearly fit the templet, using

Fig. 221. Rest Blocked Out.

the roughing gouge in the center (Figs. 20 and 78), and chisels on the balls (Figs. 23, 33, 53, and 54). Finish the piece by scraping (see Figs. 127, 128, and 211).

Fig. 222. Using a Templet.

As the scraping will tear the wood, unless the chisel is very sharp and is cutting but a very little, you will need to work carefully and keep your tools very sharp.

ELEMENTARY TURNING

If when the piece is fitted to the templet there is yet torn grain, the smoothing of it will injure the shape of the piece. Hold the templet as shown in Fig. 222. Do not press it against the work while in motion.

You must also be very careful about the sandpapering or it will change the curves so that they will not be correct. In trimming the ends you must also allow for smoothing with coarse sandpaper, or there will be a flat place made which will seriously injure the appearance of the balls.

NUMBER XXII

PIN TRAY

The making of the pin tray illustrates a method applicable to the making of many small articles.

The reason for using a spur chuck, Fig. 275, instead of a screw chuck or of gluing a piece to the chuck with paper between, is to save time and trouble.

See that the blank has been sawed nearly to size before placing it on the chuck (Fig. 224). Turn the edge and a little of

Fig. 223. Pin Tray.

each surface near the edge with the tee rest set as in Fig. 225. Use the tool as in Figs. 125 and 126.

Fig. 224. Pin Tray Blank on Chuck.

Before sandpapering the edge, set the rest as in Fig. 226, and turn nearly all of the inside, using the tools, as in Figs. 126, 127, and 128. Remove the rest, and then sandpaper and polish the edge and a little of each side.

After the edge has been polished, place the piece in a chuck, as shown in Fig. 227, and then finish the bottom. The last step is to reverse the piece in the chuck and finish the center and the inside.

Fig. 225. Pin Tray, Edge Turned.

If the edge is marred while in the cup chuck, the tray may be placed on the spur chuck, using the small holes as at first, and then refinish the edge. A block

Fig. 226. Pin Tray, Inside Turned.

ELEMENTARY TURNING 147

Fig. 227. Pin Tray in Cup Chuck.

should be placed between the end of the dead-center and the finished inside surface.

NUMBER XXIII

TURNED FRAMES

Frames, either square, as shown in Fig. 228, or round, as shown in Fig. 229, may be turned on the face-plate. They should be securely fastened by using four screws the same as in securing the wooden facing for the screw chuck, Fig. 268, the candlestick base, Fig. 195, or the blank for the molding, Fig. 259.

Sometimes frames are held only by a central screw the

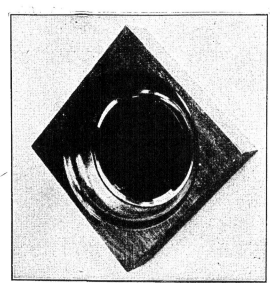

Fig. 228. Square Frame.

same as the rosette (Fig. 125). This will be sufficient to hold them, if care is exercised in doing the turning. For the first attempt you had better use the four screws.

Fig. 229. *Round Frame.*

If the frame is to be square, it should be of an even thickness before placing it on the chuck, so that there will be no need of doing more work with the turning tools than to cut the circular opening.

If the blank is so large that the screws from the holes in the iron face-plate would enter the part which is to be cut out, first fasten to the face-plate a larger wooden facing, and then secure the frame to this, as in Fig. 230.

Be very careful in placing the frame on

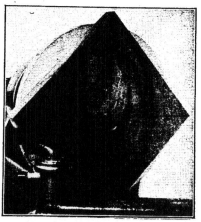

Fig. 230. *Frame Blank on Face-plate.*

ELEMENTARY TURNING 149

the chuck so that the opening can be cut from the center. A good way to center the piece is to locate the center by either of the methods shown in Fig. 2 or 3.

Place the face-plate on the spindle, and crowd the blank for the frame against it by moving the tail-screw against the center of the blank.

Fig. 231. Using Gouge on Frame.

Fig. 232. Using Gouge on Frame.

Mark the position of the blank, and after removing the face-plate from the lathe, fasten the blank to place with the screws. Fig. 230 shows the blank in place and the diagonal lines used in locating the centers. It also shows the small center made at the intersection of the lines by the end of the dead-center.

The opening should be worked with the gouge, holding it as shown in Figs. 231, 232, and 233. First hold the gouge as in Fig. 231, being careful to roll it so that it will not run towards the outside edge of the frame. The principle which governs its action is the same as in starting the cove (Fig. 47). If the cutting edge lies in the circle, it will not tend to run either way, but will cut freely and rapidly.

Fig. 233. Using Gouge on Frame.

After starting the opening at the outer edge, reverse the gouge and cut from the center, as in Fig. 232. Do not remove a larger amount of material than is needed to form the curve; the remaining waste material at the center will fall out as the gouge cuts through to the chuck.

After the opening has some depth, the gouge may be held at a greater angle, as shown in Fig. 233. If you are careful to hold the gouge properly, the curve will be shaped in a very few minutes.

Be very cautious to keep your hands and clothing away from the corners of the revolving piece.

ELEMENTARY TURNING

After the frame has been shaped with the gouge, finish the curves with the scraping tools (Figs. 102 and 128). Do not touch the surface, which is to remain flat, with any of the lathe tools.

After the center has been removed, and the curve properly finished by scraping, sandpaper the curve. Do not allow any sandpaper to touch the flat face. Remove the tee rest, and then hold the sandpaper as shown in Fig. 234. By holding one hand with the other, you will avoid the danger of your hand slipping and being hit by the corners of the frame. Entirely finish the turned parts before removing the frame from the face-plate.

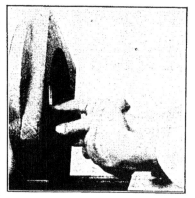

Fig. 234. *Using Sandpaper on Frame.*

The opening may be cut only a part of the way through, and the mirror or picture be held in place by using a small reed, as shown in Fig. 228; or it may be cut entirely through, and the picture or glass placed against the back, as shown in Fig. 229.

If a space at the back is desired, first secure the frame to the face-plate with screws, passing into the waste material. After the back opening has been finished, reverse the piece, and work from the face side. An easy way to center the piece for reversing

152 ELEMENTARY TURNING

is to bore a small hole through the center, and with the dead-center in this hole, force the piece to place.

After the piece has been removed from the face-plate, finish the face and edges by using plane,

Fig. 235. Square Frame, Finished.

scraper, and sandpaper, and then polish the same as the turned part. The round frame (Fig 229) is made in the same manner, except the edge which is turned the same as the rosette (Figs. 125 and 126).

ELEMENTARY TURNING

The face also may be finished in the lathe. There will then be no hand finishing to do after the frame is removed from the lathe. It is better to finish the flat surface parallel with the grain of the wood, after removing the piece from the lathe.

NUMBER XXIV

CHAIR LEGS

Fig. 236 illustrates a typical form of chair leg. The principles involved in turning chair legs do not differ any from those learned in turning the first

Fig. 236. Square-topped Chair Leg

twenty exercises. Fig. 237 shows the general arrangement of the lathe for turning long work, and also the position of the hands and the body.

In roughing long pieces, it is usually best to begin near the dead-center and turn down but a little at a time, as shown in Fig. 238. Each time begin a little farther to the left, and finish a little of the cylinder at the right.

Fig. 239 shows the method of working the pattern. First, turn the piece to the general outline, then

Fig. 237. Position in Roughing Long Piece.

ELEMENTARY TURNING 155

begin at the end, usually at the top end, and work out the pattern. The completed leg is shown in Fig. 236.

Fig. 240 shows a design in which there is a square part to receive the rungs. For such legs the rungs or rails should be of rectangular section.

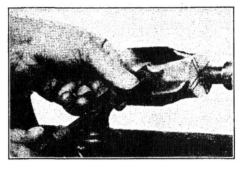

Fig. 238. *Roughing Gouge on Long Piece.*

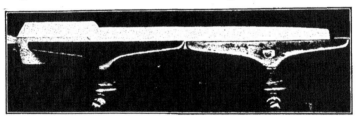

Fig. 239. *Chair Leg Outlined.*

Fig. 240. *Chair Leg with Square Section.*

Fig. 241. *Chair Leg.*

Fig. 241 shows a typical form of leg for use in wood bottom chairs. The number of rungs used must be considered in determining the pattern.

NUMBER XXV

CHAIR RUNGS AND SPINDLES

The lighter parts of chairs, such as rungs or stretchers and spindles are somewhat difficult to turn, because they spring so easily. To avoid the

Fig. 242. *Chair Rung.*

springing, first turn the piece to a cylinder the entire length (Fig. 243), and then turn the center of the piece as shown in Fig. 244; the piece may be steadied by the hand as shown in Fig. 183.

Fig. 243. *Chair Rung Roughed to a Cylinder.*

Finish the ends, turning the parts for the tenons to near the finished size. Hold the sizer as shown in Fig. 245, and size the tenons. If the piece tapers

Fig. 244. *Chair Rung Center Turned.*

to the tenon, it will be necessary to finish down to the tenon with the skew chisel after using the sizer.

ELEMENTARY TURNING

If the live-center is larger than the finished size of the tenon, there will be a small stub as shown in Fig. 242. This must be split off with a knife after the piece has been removed from the lathe.

In planning the design for a rung or spindle, be very particular to avoid any deep cuts near the center.

Fig. 245. Using the Sizer.

Fig. 246. Spindle.

Fig. 247. Plain Spindle

Figs. 246 and 247 show two styles of spindles. The same general plan is followed in turning them as in turning the legs and the rungs of a chair.

NUMBER XXVI

FOOTSTOOL LEGS

In designing footstool legs as in all other designing, try to have a fair idea of the form which you wish to make before commencing to shape the material.

158　ELEMENTARY TURNING

You should, if possible, have a sufficiently definite idea of the design you are to make to allow of first outlining the piece, as shown in Fig. 249.

Fig. 248. Foot tool Leg. Size of stock 1¾ inches square by 8 inches long.

The next step is to turn each end as shown in Fig.

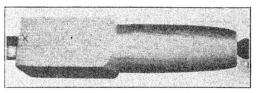

Fig. 249. First Step in Turning Footstool Leg.

250. This method will help you to proportion the various parts.

Fig. 250. Second Step in Turning Footstool Leg.

Finish by turning the long curve, completing the design, as shown in Fig. 248.

NUMBER XXVII

DESIGNS FOR FOOTSTOOL LEGS

Figs. 251, 252, and 253 suggest a variety of patterns for footstool legs. See also Figs. 236, 240, 248, and 254.

ELEMENTARY TURNING

Do not attempt to copy any of them, but study carefully each one, and then work out a design of your own. Remember that good designs are usually simple, and that a few elements properly combined are far better than many carelessly brought together.

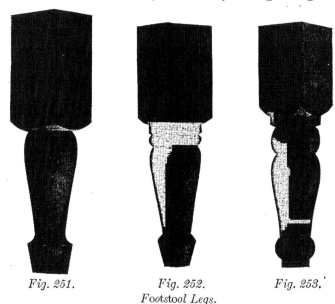

Fig. 251.   Fig. 252.   Fig. 253.
Footstool Legs.

One of the facts most difficult to realize in turning is, that a very slight change in a curve, or in the proportion of parts will change a piece from ugliness to beauty. In working out a design at first use a wood easily turned. Pine or basswood is probably the best wood to use. You can usually begin at the end nearest the live-center and make this end the top end of the leg.

160 ELEMENTARY TURNING

Unless there is some good reason for doing differently, follow the general plan described in turning Fig. 248.

Unless you are more successful than most turners, you will need to try several times before making a good original design. All designing of similar shaped pieces follows this general plan. After you have completed the design and know exactly what shape you wish, you can plan such an order in the use of tools as will result in the greatest speed.

NUMBER XXVIII

FOOTSTOOL

Fig. 254 illustrates a footstool completed, except the upholstering. The length of the side rails as well as the height of the legs may be changed to suit individual requirements.

Fig. 254. Footstool.

Figs. 248 to 253 show various styles of legs which may be used for footstools.

ELEMENTARY TURNING

The dimensions of this stool are: Legs, 2½ inches square by 12 inches long; side rails, ⅞ inch by 3½ inches by 12 inches between the legs.

The ordinary turning stock, 1¾ inches square, is large enough for most footstool legs. The length may be anything from 8 to 16 inches. Footstools should not be higher than they are wide.

NUMBER XXIX

TURNED PIANO STOOL

The usual sizes for such a stool are: top, 14 inches in diameter by 1¾ inches thick; legs, 1¾ inches in diameter by 19 inches long, to the under side of the top. As they enter the top 1 inch, the stock should be 20 inches long. The rungs are made from ⅞-inch square stock and are 10 inches below the

Fig. 255. Piano Stool.

under side of the top. The distance between the legs at the rungs is 8 inches, making the total length of the rungs 10 inches. The holes in the top for the ends of the legs are bored in an 11-inch circle with a $\frac{7}{8}$-inch bit. This should make the diagonal distance on the floor between the centers of the legs about 16 inches.

First, turn a pattern for a leg in soft wood or a cull piece. If you wish to make a very fine piece of furniture, polish every part in the lathe.

In fitting the parts together be very careful not to injure the finish. Blocks of soft wood, shaped to fit the turning, will aid much in holding the legs while boring the holes for the rungs. Study the design carefully and see if you can improve upon it.

NUMBER XXX

TURNED STOOL

Fig. 256. Turned Stool.

The top of this stool is 14 inches in diameter by $1\frac{3}{4}$ inches thick. The legs are but 18 inches long, which includes the 1-inch tenon entering the top. The crosspieces are $10\frac{1}{2}$ inches above the floor, and the legs are $10\frac{1}{2}$ inches apart from surface to surface at this point.

ELEMENTARY TURNING 163

The holes in the top are bored in a 10½-inch circle. Read the description of No. 255, and compare it with this stool, and then work out a new design.

NUMBER XXXI

GROUP OF FANCY TURNINGS

These designs may be used for suggestions or in place of those given in the text.

Fig. 257. Group of Fancy Turnings.

NUMBER XXXII

TURNED MOLDING

Sometimes it is necessary to make circular pieces of molding to be used at rounded corners or at semi-circular ends. This is done by turning a complete circle, and then cutting from it such segments as are required.

Fig. 258 Turned Molding.

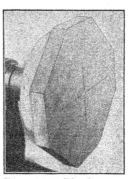
Fig. 259. Blank for Turned Molding.

Fig. 259 shows a piece of ⅞-inch board secured to a face-plate. No attempt has been made to make the blank a true circle before placing it in the lathe, for the rough corners of this octagonal shaped piece of soft wood can be cut away quite easily in the lathe. As the diameter of the circle is greater than the

Fig. 260. Molding Segments.

ELEMENTARY TURNING

diameter of the iron face-plate, a wooden facing is attached, so that the screws will enter the part of the wood that is to become the molding. If the screws were to enter the waste material, you could not finish the inner edge of the molding.

Adjust the rest, and turn the outer edge as in working the rosette (Lesson 30). Turn the inside of the circle, as in making the frame (Fig. 235).

After the molding has been finished, as in Fig. 258, it may be cut into such pieces as are required.

Fig. 260 shows one section for a rounded end, one for a rounded corner, and one to connect parts at an angle of sixty degrees.

TOOLS AND FITTINGS

PART III

INTRODUCTION

This part describes such tools and materials as are required for the work in this course, except such tools as the pupil has become familiar with in his use of the author's previous publication, entitled Elementary Woodwork.

Work at the bench in every case should precede the study of turning. There are many points about the use of tools which must be understood in order to do turning properly, and which can be learned much better by working at the bench.

The number of tools described is the minimum rather than the maximum number which may be used in wood turning. For doing the work on supplementary and fancy pieces, many special tools might be used. The use of these tools, however, requires no additional instruction, as they are only modifications in form of those described and used in the various examples which have been given. One who has executed all of the models illustrated in Parts 1 and 2 should understand how to proceed in turning any but the most difficult work.

168 ELEMENTARY TURNING

The materials for finishing are properly limited to the least number possible. Those who desire a larger variety of finishes should consult a work devoted entirely to finishing and polishing.

ARBORS

For the ordinary work of a wood-turning lathe only the simple wooden arbor is required. Such arbors are shown in use in Figs. 163, 172, 174, and 206. They should be made of bits of waste material. Usually they are of hard wood, yet for such uses as holding of rings for re-polishing, soft wood may be used.

Be sure to force the arbor on to both centers far enough to hold it from slipping on the live-center. The marks at each end should be large enough to permit the arbor to be removed and replaced in exactly the same position. Do not forget to mark it as indicated in Figs. 9 and 10.

In fitting the arbor to the work, make the taper so slight that there will be a firm bearing nearly the whole length of the hole. Avoid jamming the ends in forcing the work either on or off of the arbor Arbors should be carefully made, and kept for future use.

CALIPERS

Fig. 261 illustrates one form of outside calipers, and Figs. 58 and 132 show how the outside calipers

ELEMENTARY TURNING 169

are held. Fig. 262 illustrates inside calipers. Fig. 263 shows how the inside calipers are held.

There are many styles of these very useful tools. The chief difference is that some have a device for fine adjustment, and others do not. Those having a screw adjustment are better for the beginner to use.

Fig. 261. Outside Calipers.

Fig. 262. Inside Calipers.

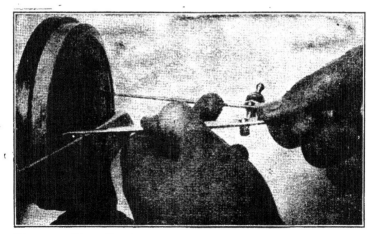

Fig. 263. Using Inside Calipers.

ELEMENTARY TURNING

Fig. 264. Caliper Ends.

If the points are sharp, they may catch in the wood, and, therefore, they should be rounded, as shown in Fig. 264. If the ends are not rounding, do not attempt to use them while the lathe is in motion.

CHISELS

The turner's chisel, called the turner's skew chisel, and shown in Fig. 265, is used on nearly every piece of work. It differs from the common firmer

Fig. 265. Skew Chisel.

chisel by having a bevel at each side, and the cutting edge at an angle. Fig. 266 indicates the shape of the cutting edge.

Turning chisels are much heavier than the common bench chisels; and there is no bolster at the end of the handle to prevent it being driven too far on to the chisel. Turning chisels are usually made in sizes from $\frac{1}{4}$ inch to 2 inches, but for ordinary light work a $\frac{1}{4}$-inch, a $\frac{1}{2}$-inch, and a 1-inch chisel are sufficient. The shape and angle

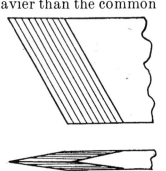

Fig. 266. Cutting End of Skew Chisel.

ELEMENTARY TURNING 171

of the cutting edge has much to do with the ease with which the chisel is used.

Great care should be taken in grinding and whetting these tools. Unless the grindstone is quite true, it is useless to attempt to grind such chisels, except by holding them free-hand, as shown in Fig. 267.

Fig. 267. Grinding a Skew Chisel.

To hold them free-hand is not difficult. One hand bears them against the stone, and the other hand holds the blade at the proper angle.

This is done by a combination of two movements. The hand may be raised or lowered, or it may revolve the chisel handle.

By watching the flow of water past the tool, and by frequently removing it to see where the stone is cutting, you may soon learn how to grind a skew chisel properly.

Do not be satisfied until the cutting edge is straight from the acute to the obtuse angle, and both angles of a correct size. The two bevelled surfaces should also be perfect. If there is any roundness near the edge, it will hinder the chisel from resting properly on the work, and render it much more liable to catch and injure the piece.

In whetting the skew chisel, keep the surface as near true as possible. Do not form another angle with the oilstone, as in whetting the carpenter's chisel. The skew chisel is for cutting, and should not be used as a scraping tool. This, however, is done sometimes because no other chisel is at hand. Chisels for scraping are called scraping tools, and are described under that head.

CHUCKS

There is scarcely any limit to the number of styles and sizes of chucks which can be used for wood turning. For elementary work only a few are required.

The screw chuck, as shown in Fig. 268, is the one most often used. The spur chuck (Fig. 275) is very handy for some work. Other styles, except the wooden cup chucks which are often used as shown in Figs. 165, 168, 169, and 227, may be dispensed

ELEMENTARY TURNING 173

with for all ordinary work. The cup chucks are very simple appliances, being merely blocks of wood, secured to a face-plate, and hollowed out to receive the work.

SCREW CHUCK

There are many kinds of screw chucks. The beginner can easily do all his work with the common style, such as is shown in Fig. 268. This chuck is made by attaching a piece of wood to the ordinary face-plate and securing a common wood screw at the center, as shown in Fig. 274. Hard wood is better for the facing.

It should be securely fastened with at least four screws to the iron face-plate, and turned smooth at the circumference, so that the hands will not be liable to be injured by it. Do not use any sandpaper on it. Figs. 269 and 270 show how the gouge is held in turning the edge. Figs. 125 and 126 show the gouge in use for similar work. The character of the piece will usually determine whether the gouge should first cut from the right or from the left.

Fig. 268. Screw Chuck.

174   ELEMENTARY TURNING

Turn the face of the piece exactly true, testing it with a straight edge. The roughing gouge should be used at first, and held as shown in Fig. 271.

Fig. 269. Turning an Edge with Gouge.

The final smoothing should be done with a scraping chisel, as shown in Fig. 127. If no scraping chisel is at hand, a skew chisel may be used for the scraping. Unless the face is exactly straight, the blocks when screwed against it may not remain in place.

Find the center, as shown in Fig. 272. By placing the point of the skew chisel near the center, a small circle will be made; and by gradually moving the point towards the center of the circle, a place

Fig. 270. Turning an Edge with Gouge.

will be found where no circle is made. This will be the center, and the point of the chisel should make a small hole in which to place the spur of the bit. Bore a hole just large enough to fit the shank of the screw. Use the method shown in Fig. 273. Place the bit, and then carefully start the lathe. Mark the piece so that you can put it back in the same place, and then remove it from the iron face-plate.

Fig. 271. *Turning the Face with Roughing Gouge.*

Countersink a place just deep enough to bring the surface of the screw head flush with the wood.

Drive a finishing nail into the wood at the end of the slot in the screw head and bend it over, as shown in Fig. 274. If the nail is too large to go into the slot, hammer it flat near the center before attempting to drive it.

Fig. 272. *Finding the Center.*

Return the wood to its place on the iron plate, and you have

176 ELEMENTARY TURNING

the best kind of a screw chuck for general work. Be careful to turn each screw until it is just tight; for, as they may be several times removed and again

Fig. 273. Boring with Bit.

Fig. 274. Back Side of Wooden Facing.

inserted, any overstraining will spoil the holes in the wood.

If you are careful to put grease in the holes each time the screws are to be inserted, the wood will not wear out so quickly, and the screws will hold stronger.

ELEMENTARY TURNING

SPUR CHUCK

Pieces of some shapes that cannot be screwed on to the screw chuck can be held on a spur chuck (Fig. 275). This chuck is simply a wooden facing, having two or more short spurs in its face side and secured to an iron face-plate. The piece to be worked is forced on to these spurs, and held against them by the tail-stock, as shown in Fig. 224. While in this position, the edge and nearly all of one side can be finished. The spurs do not usually project more than $\frac{1}{8}$ inch, and $\frac{1}{16}$ inch is sometimes sufficient. They may be of any desired number, and should be firmly driven into the facing, so that they will not be pressed deeper into the facing, when pieces to be turned are forced against them.

Fig. 275. Spur Chuck.

The spurs may be made of wire nails by driving the nails into the facing from the face side, and then cutting them off and filing them sharp. For heavy work the nails may be driven in from the back side, and their heads left to rest against the iron face-plate, so that forcing pieces on to their sharpened ends will not move them.

COMPASSES

The plain compasses, shown in Fig. 276, are sufficient for much of the spacing work in turning.

At times the wing compasses, used by joiners, are better. If one set is to be used for both joinery

Fig. 276. Plain Compasses.

and turning, those having the fine adjustment should be procured.

The use of the compasses in marking off spaces is shown in Fig. 81.

DEAD-CENTER

The dead-center should be of hardened steel and shaped as shown in Fig. 277.

The central point should be removable so that it may readily be replaced if broken. The rim and

Fig. 277. Dead-center.

cup surface should be very smooth and bright. It should be forced to place at the same time the blank is forced on to the live-center, and therefore, no hammer or mallet should be used in placing the work in the lathe.

ELEMENTARY TURNING

FACE-PLATES

Face-plates are usually of the form shown in Fig. 278. See also Fig. 269. These screw on to the outside of the end of the lathe spindle.

Fig. 279 illustrates a style which is fastened to a plug, which is tapered to fit the hole in the end of the live spindle. For small work the

Fig. 278. Face-plate.

latter style is better, because it allows the screws to be arranged on a smaller circle. Such an arrangement is not so strong as the first form, but it is quite essential for some work.

Fig. 279. Small Face-plate.

GAUGES

For determining the horizontal distances on work in the lathe, many forms of gauges and templets are used. For most work, when but a few pieces of a kind are to be turned, a rule and a pencil

Fig. 280. Gauge.

(Fig. 19) or a rule and a chisel point (Fig. 31) are sufficient.

One of the most common forms of gauges for use in turning is shown in Fig. 280 and in use in Fig. 36.

These gauges may be made of various styles, either from the drawings or from the model piece.

The edge of the stick or bar may correspond

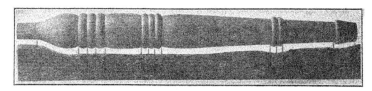

Fig. 281. Chair Leg Gauge.

with the general outline of the pattern, as shown in Fig. 281.

After the bar has been shaped, drive brads or nails into the edge. Be careful to place each one exactly opposite the point which is to be lined. Cut off the heads and adjust each one to length.

By making the spurs of the correct length they may help to indicate the size, as their points may be made to cut each time to the same depth. File the spurs to a wedge-shaped point, being careful that the point is in exactly the correct place. The points may be sprung with a hammer, but are liable to gradually spring back. See that they are driven well into the wood, for the revolving of the work against them tends to move them.

ELEMENTARY TURNING 181

Unless there are many pieces of a kind to be turned, it is better to use the pencil and rule, or compasses rather than to take the time to make a gauge of this kind.

Often a wooden tee rest is used and the pattern marked on its edge so that no other measuring is required.

GOUGES

Fig. 282 illustrates an ordinary turning gouge. The curve at the cutting edge varies greatly for use in different kinds of work. The end is shown from different angles in Figs. 44, 47, 126, and 232.

Fig. 282. Turning Gouge.

Figs. 40 to 44, 53, 92, 111, 114, 116, 120, 125, 126, 170 show the turning gouge in use.

Turning gouges may be ground square across and used for roughing, but usually a firmer gouge (Fig. 283) is used for this purpose. For ordinary

Fig. 283. Firmer Gouge.

work the firmer gouge is strong enough; and because the metal is thinner, it is much easier to keep it in order.

182   ELEMENTARY TURNING

The roughing gouge is shown in use in Figs. 13, 20, 78, 100, 113, 238, and 271. The grinding of gouges is not an easy task. Their cutting edges should be free from all roughness, and their curves regular throughout.

In order to produce such an even edge the gouge should be held on the grindstone as shown in Fig.

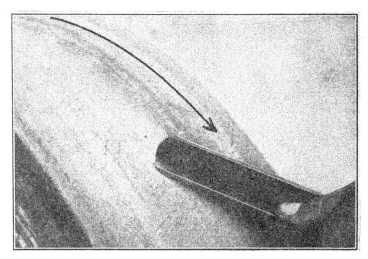

Fig. 284. Grinding a Gouge.

284. The angle of the basil side is determined by the position of the right hand. By rolling the hand, the grindstone is made to cut at any part of the curve. Do not attempt to grind the tool to an edge at one point, and then roll it a little in order to grind at another place, but rather keep the gouge rolling from one edge to the other edge.

ELEMENTARY TURNING 183

In grinding the turning gouges, the handles will need to be lowered and raised as well as rolled. If there are any thick places that require more grinding than others, do not stop the gouge at these places, but rather do not roll it so rapidly. This carefully done, will result in a very even curve.

Fig. 285. Whetting a Gouge.

The rapid passing of the shavings over the cutting edge wears it quite rapidly, and therefore the inside as well as the basil side of the gouge is worn. This makes necessary the grinding back of the edge a little in order to make the inside edge straight.

Fig. 286. Using a Slip Stone.

After the gouges have been ground, their edges should be smoothed by rubbing an oilstone on them, as shown in Fig. 285.

The stone may be held on the bench and the gouge rubbed on the stone. Whichever method is employed, the movement must be such that the stone moves along the edge at every stroke, so that the edge is sharpened evenly.

To remove the wire edge, a hard Arkansas slip stone is used, as shown in Fig. 286. This stone should be rubbed towards the edge at the same time it is given a side movement, causing it to come in contact with the entire cutting edge. This stone may be used on the basil side as well as the straight side. This stone should produce so fine an edge that no leather strop will be required. Remember that in whetting turning tools, the changing of the angle soon necessitates regrinding. By being careful about the whetting, you can save much time in the grinding.

LATHES

Fig. 287 illustrates a modern all-metal lathe for wood turning. Such lathes may be adapted to the working of brass and soft metals.

In elementary turning, all you have to deal with is the adjustment of the tee rest and the tail-stock; the changing of the belt to a suitable speed, and the exchanging of the live-center for face-plate, or vice versa.

Your lathe may not be exactly like the one shown in the picture, yet all woodworking lathes for hand turning are similar. The one shown in Fig. 288 is

ELEMENTARY TURNING 185

quite equal to an all-metal lathe for many kinds of work. By comparing the two you will notice that the cones are not placed alike, and also that the oil holes are not in the same position. Fig. 288 shows the parts of the lathe except the wooden bed or

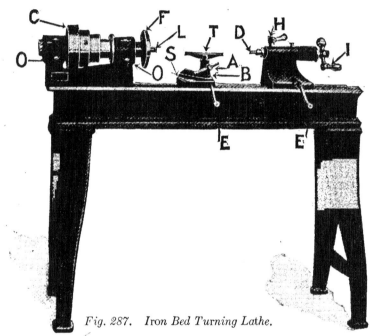

Fig. 287. Iron Bed Turning Lathe.

A Set Screw. B Tee Rest Stand. C Cone Pulley. D Dead-center.
E Clamp Handles. F Face-plate. H Clamp Screw. I Crank Handles.
L Live-center. O Oil Holes. S Shoe. T Tee Rest.

shears. The long bolts are for securing the headstock, tail-stock, and rests to the bed. Two tee-rest stands are shown. There is also a double tee rest for long work. In elementary turning there is so

186 ELEMENTARY TURNING

seldom need for a long rest that usually none is provided and instead, two short ones are set side by side as shown in Figs. 239 and 243. The lathe shown in Fig. 287 has an oil cavity under the bearing, so arranged that the oil is fed up to the bearing as required, and hence a quantity of oil is put into the reservoir once in a long time.

Fig. 288. Wooden Bed Turning Lathe.

Most lathes are oiled as shown in Fig. 288, and require oiling once or twice for every ten hours of running. Watch your lathe bearings by touching them with your fingers, and if they are hot, report it to the instructor. Sometimes all that is required to cool them is a few drops of oil, but at other times it is necessary to adjust the boxes. You should not attempt to make any adjustments of the bearings unless you are sure that you will do no harm.

ELEMENTARY TURNING

OILSTONES

The oilstones used for sharpening lathe tools do not differ from those used in sharpening joiners' tools. Both should cut freely and smoothly. If the tools are rubbed on a strop after whetting, the edge will be made smooth, and it will cut better.

Gouges are not easily rubbed on a strop, and, therefore, a very hard stone is used to remove the wire edge, and give the smoothness re-
Fig. 289. Oilstone Slip.
quired for fine work. Such stones are usually shaped as shown in Fig. 289, and are called oilstone slips. They vary greatly in size. For sharpening ordinary turning tools, use a slip about 3 inches long by $1\frac{1}{2}$ inches wide.

The grade known as hard Arkansas is excellent for school use. These stones are very brittle, and should be used with care. Use oil on them the same as on the India oilstone, used in the wood shop.

PARTING TOOLS

Fig. 290. Parting Tool.

The parting tool, shown in Fig. 290 and in use in Figs. 106, 110, and 122, is for cutting deep recesses or for cutting pieces in two. Its work is always

188 ELEMENTARY TURNING

rough, consequently its use is limited. It cuts rapidly, but always roughly, and therefore saves no time, except when the surface being cut is not required to be smooth.

It should be held with the point or cutting edge directed towards the line of the lathe centers. Usually it should be made to cut a space wider than itself by being moved a little from side to side.

SCRAPING TOOLS

Fig. 291 shows an ordinary right-hand scraping tool which was made from a worn-out turning chisel.

Fig. 291. Scraping Tool.

Fig. 292 shows the shapes of the ends of the ordinary scraping tools. They may be of any shape required to fit the work. They may also be crooked or bent to reach into obscure interior curves. They should never be used when a cutting tool can be used. They are for finishing, and not for removing large amounts of material.

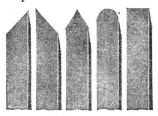

Fig. 292. Scraping Tool Ends.

Scraping tools are usually held level with the lathe centers, as shown in Figs. 127, 128, 198, 211, and 207. They are sharpened much the same as ordinary firmer chisels. It is not necessary that the cutting angle be as small as the angle on firmer chisels. For making the finishing cuts they must be very sharp.

Scraping tools are properly pattern-makers' tools, and should be seldom used in cabinet turning. Scraping tools, including ring tools and similar special tools, are usually made from worn firmer chisels, or short turning tools.

RING TOOLS

The ring tools shown in use in Fig. 198 are special forms of scraping tools. The hook-like ends

Fig. 293. *Right-hand Ring Tool.*

Fig. 294. *Left-hand Ring Tool.*

may be formed by grinding firmer chisels, as shown in Figs. 293 and 294. Fig. 295 shows the position of the tools while in use. A is a section of a part of the arbor, and R is a section of the ring, and T the ends of the tools. As these are light tools, they should be used carefully.

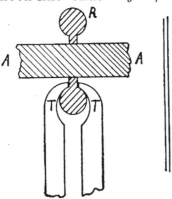

Fig. 295. Section Showing Cutting Position of Ring Tools.

SIZING TOOLS

The turner's sizing tool shown in Fig. 296 is used for sizing tenons in the lathe. It is held as shown in Fig. 245. The place to be sized should be of nearly the desired dimension before applying the tool. This is not an easy tool to use and must

Fig. 296. Sizing Tool.

be held very firmly or it will cut too rapidly or catch and injure the work.

The adjustment is made by loosening the thumb-screw and moving the hooked-shaped piece until the opening at the cutting point is of the correct size. After adjusting, it should be tested on a piece of waste material for it often cuts smaller than is expected.

SPUR CENTER

The spur center is a very important part of the lathe. It should be shaped so it will enter the wood easily and hold securely. The one shown in Fig. 297 is a good design.

Fig. 297. Spur Center.

The center point should be a separate piece and should be easily removed for sharpening or replacing.

ELEMENTARY TURNING 191

The point should be of steel, but not necessarily tempered.

In placing the spur center in position in the arbor, do not drive it with a hammer or a mallet, for the pressure against it in placing the wood between the centers will be sufficient to force it tight enough to not slip while in use.

The spur center should be removed by driving a key, made for this purpose, into a hole in the side of the arbor and against the end of the center. Some centers have a square part and are removed by using a wrench.

Some lathes are supplied with several spur centers, but for all ordinary turning, one is sufficient.

TEMPLETS

Templets are thin pieces of wood or metal, so formed as to determine the outline of another

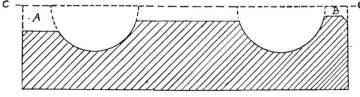

Fig. 298. Templet.

piece. The one shown in Fig. 298 and in use in Fig. 222 represents the most common class. They may be of almost any size or shape for work within the capacity of the lathe.

192   ELEMENTARY TURNING

Fig. 298 shows a method of laying out a templet. First, draw line C——C, representing the axis of the piece, from this lay out the shape required.

After the outline has been determined, you must decide upon the size of the stub to be left at each end and cut off from each end of the templet an amount equal to one-half the diameters of the stubs. This is shown by the dotted lines at A and B. The large stub, A, is at the live-center.

Be careful in drawing the lines, and work the templet carefully to shape, as you are not likely to succeed in fitting the piece to it exactly. If the templet is not quite correct, your piece may be very much out of shape.

In using templets, do not hold them against the work while it is in motion; for if you do, they will very soon become incorrect.

INDEX

Arbor	119–123–124–135–138
Back Screw	127
Base for a Candlestick	134
Base for a Hat Rest	139
Beaded Spindle	69
Bead and Cove	58–61
Beads, Turning	46–60–63–70–80–85–98–130
Blank on Face-plate	139–148–164–174
Blank on Screw Chuck	87–94–101–121–164
Boring with Bit	111–176
Boring with Gouge	87–95–123–131
Box	86–92–132–142
Calipers	63–107–169
Candlestick	133–136
Carpenter's Mallet	113
Carver's Mallet	115
Centering Material	19–20–146–148–164–175
Chair Leg	153–155
Chair Rung	156
Chair Spindle	157
Chisel (See Skew Chisel)	
Chuck	120–122–146–147–172
Clamp Screw	22–185
Cleaning Lathe	16
Combining Woods	141
Compasses	79–178
Cover	90–93–129–133
Coves	51–53–55
Cup Chuck	120–122–147–173
Curved Spindle	81

ELEMENTARY TURNING

Curves, Turning	36–42–46–60
Cutting in for Square Ends	76
Cutting Threads	128
Cutting with Point of Skew Chisel	44–51–76–89–91–120
Cylinder	25
Darning Ball	117
Darning Hemisphere	118
Dead-center	21–146–178
Designs	
Box	86–92–129–132–142–163
Candlestick	134–136–163
Chair Leg	153–155
Footstool Leg	158
Frame	147–152–163
Gavel	109–112
Goblet	94–97–143–163
Hat Rest	137
Mallet	112–114–115
Napkin Ring	121–124
Equipment	14
Examining Work	25
Face-plate	179
Finding the Center	175
Footstool	160
Footstool Leg	157–159
Frame	147–148–152
Gauge Stick	49–179
Gavel	109–112
Glove Mender	118
Goblet	93–97–142–143
Gouge	51 to 60–83–87–99–102–123–131–149–174–181
Gouge, Grinding and Whetting	182
Grinding	171–182

ELEMENTARY TURNING 195

Group of Fancy Turnings	163
Half-inch Bead	47
Half-inch Cove	57
Half-inch Bead and Cove	61
Half-inch Left-hand Semi-bead	41
Half-inch Right-hand Semi-bead	44
Handle	106–108–112–125
Hand-screw Screws	127
Hat Rest	137
Illustrations, Remarks on	16
Inside Calipers	169
Introduction to Part I	11
Introduction to Part II	104
Introduction to Part III	167
Jointing Surfaces for Gluing	138
Knife and Fork Rest	144
Knob, Turning	93–126
Large Box	129
Lathe	185
Leather Topped Handle	108
Left-hand Semi-bead	35
Mallet	112–114–115
Marking for Replacing	24
Marking Spaces	32–44–76
Molder's Rammer	116
Molding Segments	164
Napkin Ring	121–124
Oiling	15–21–186
Oilstone	187
One-inch Bead	46
One-inch Cove	50

One-inch Bead and Cove – – – – – – 58
Outside Calipers – – – – – – 63–107–169

Parting Tool – – – – – – – – 187
Personal Equipment – – – – – – – 14
Piano Stool – – – – – – – – 161
Pin Tray – – – – – – – – 145
Placing Shellac on Cloth – – – – – – 72
Placing Work in the Lathe – – – – – 19–21
Polishing – – – – – – 71–96–110–123
Polishing Outfit – – – – – – – 74
Porch Spindle – – – – – – – – 84
Pumice Stone – – – – – – – 69

Rebate – – – – – – – – 88–91
Regulations – – – – – – – – 15
Rest Inside of Bowl – – – – – – – 95
Right-hand Semi-bead – – – – – – 37
Ring – – – – – – – 98–119–135
Ring Tools – – – – – – – 135–189
Rolling the Gouge – – – – – – 33–78–83
Rosette – – – – – – – – 101
Roughing Gouge – – – – – – – 181
Roughing Gouge, Use of– 26 to 33–78–154–175
Rounding a Blank 27–78–87–94–101–146–154–174
Rounding Corners – – – – – – – 77
Round Frame – – – – – – – 148
Rungs – – – – – – – – 156

Sandpapering – – – – – – – 65–95–151
Scraping Tools – – – – – 102–103–140–188
Screw Box – – – – – – – – 128
Screw Chuck – – – – – – 101–121–173
Shellacing (See also Polishing) – – – – 67–71
Shoulder Screw – – – – – – – 127
Sizer – – – – – – – – 157–190
Skew Chisel – – 30 to 50–60–89–120–123–127–130–138–170–175

Slip Stone	183
Socket Chisel Handle	108
Spacing	32–44–49–79–85
Speed of Lathe	24
Spindles	157
Spindle with Cones	62
Spur Center	190
Spur Chuck	146–177
Square Frame	147–152–163
Square-end Spindle	75
Stepped Cylinder	32
Stock for Turning	14
Stool (See also Footstool)	162
Stopping the Lathe	25
Tail Screw	21–185
Tapered Spindle	82
Tee Rest	22–185
Templet	144–191
Testing the Surface	25
Threading a Wooden Screw	128
Three-eighths-inch Bead	49
Three-eighths-inch Cove	58
Three-fourths-inch Cove	55
Tightening the Tail Screw	21
Tool Handles	105–108
Tray	145
Turned Moulding	163
Turned Piano Stool	161
Turned Stool	162
Turning Gouge	51 to 56–60–83–87–93–95–97–99–102–123–131–149–174–181
Vise Handle	125
Whetting a Gouge	183
Wooden Screw	127

Printed in Great Britain
by Amazon